THE LIQUID PHASE

THE WYKEHAM SCIENCE SERIES

The aim of the Wykeham Science Series is to introduce the present state of the many fields of study within science to students approaching or starting their careers in University, Polytechnic, or College of Technology. Each book seeks to reinforce the link between school and higher education, and the main author, a distinguished worker or teacher in the field, is assisted by an experienced sixth form schoolmaster.

THE LIQUID PHASE

D. H. Trevena

University of Wales, Aberystwyth

WYKEHAM PUBLICATIONS (LONDON) LTD
(A MEMBER OF THE TAYLOR & FRANCIS GROUP)
LONDON AND WINCHESTER
1975

First published 1975 by Wykeham Publications (London) Ltd.

Cover illustration—An instantaneous 'snapshot' of a simulated two-dimensional 'liquid'.

ISBN 0 85109 031 1

Printed in Great Britain by Taylor & Francis (Printers) Ltd.
Rankine Road, Basingstoke, Hants.

Distribution and Representation :

UNITED KINGDOM, EUROPE AND AFRICA
Chapman & Hall Ltd. (a member of Associated Book Publishers Ltd.), 11 North Way, Andover, Hampshire.

WESTERN HEMISPHERE
Springer-Verlag New York Inc., 175 Fifth Avenue, New York, New York 10010.

AUSTRALIA, NEW ZEALAND AND FAR EAST (EXCLUDING JAPAN
Australia and New Zealand Book Co. Pty. Ltd.,
P.O. Box 459, Brookvale, N.S.W. 2100.

INDIA, BANGLADASH, SRI LANKA AND BURMA
Arnold-Heinemann Publishers (India) Pvt. Ltd.,
AB-9, First Floor, Safjardang Enclave, New Delhi 11016

GREECE, TURKEY, THE MIDDLE EAST (EXCLUDING ISRAEL)
Anthony Rudkin, The Old School, Speen, Aylesbury, Buckinghamshire HP17 0SL.

ALL OTHER TERRITORIES
Taylor & Francis Ltd., 10–14 Macklin Street, London, WC2B 5N

PREFACE

F the three phases of matter the intermediate liquid phase has
roved the most difficult to unravel and it is only in recent years that
has become a feasible proposition to write a book such as the present
ne. It is hoped that this book will be of use in undergraduate
ourses on the structure and properties of matter and also as back-
round reading for sixth form students. The emphasis, wherever
ossible throughout the book, has been on the molecular approach.
fter a brief introductory chapter the radial distribution function is
troduced in Chapter 2 and the gas-like nature of liquids is discussed
Chapter 3. In Chapter 4, intermolecular forces and the
2–6 potential are considered and Chapter 5 deals with molecular
ynamics and Monte Carlo calculations, Bernal's ball models and
arious diffraction methods of studying liquid structure. Phase
hanges and the surface layer are discussed in Chapters 6 and 7,
hilst Chapter 8 is almost entirely devoted to cavitation and the
ehaviour of liquids under tension. The last two chapters deal with
e flow properties of liquids: Chapter 9 is concerned with Newtonian
quids whilst Chapter 10 deals with the rheological behaviour of
me-independent, time-dependent and viscoelastic liquids. The
athematics in the book has been kept to a minimum.

I am grateful to several friends and colleagues for their help during
e preparation of this book. First I wish to thank my friend and
hoolmaster colleague, Mr. R. J. Cooke, for the careful way in which
e read all the chapters and for making so many constructive
ggestions. Secondly I wish to express my gratitude to Professor
. N. V. Temperley of University College, Swansea, for reading a
rge part of the manuscript and for suggesting various improvements.
rofessor Temperley is a leading authority in the field of Liquid
tate Physics and my association with him goes back to 1950 when
first did research in the theory of liquids under his direction at
ambridge. I owe much to the inspiration, encouragement and
iendship that he has given me over the years. Thirdly I wish to
ank colleagues in the Physics Department at Aberystwyth:
r. W. M. Jones, Dr. D. H. Edwards and Mr. Adrian Walters read
arious parts of the manuscript and gave helpful comments while
rs. Eileen Pryce typed the whole manuscript with a competence and

efficiency for which I have nothing but praise. Fourthly I wish to thank Dr. A. A. Collyer, Sheffield Polytechnic for his assistance with Chapter 10 and for providing me with some photographs. Throughout the writing of this book it has been a pleasure to cooperate with various members of Wykeham Publications Ltd.

My interest in the physics of liquids began in 1947 when I started as a research student at Aberystwyth under the late Professor R. M. Davies. By his fine personal qualities he earned the gratitude of generations of students for the great personal interest which he always showed in them and to his memory this book is dedicated with gratitude and affection.

The Physics Department,
University of Wales,
Aberystwyth,
Wales.

D. H. TREVENA

SYMBOLS AND UNITS

THOSE symbols which are in frequent use throughout this book are listed below. Where the same symbol is used to represent more than one quantity the appropriate chapter numbers are given in brackets. Other symbols, which have limited use only, are defined in the text where they occur. Except where indicated, the subscripts l, v, c and r refer to values for the liquid and vapour phases and at the critical and triple points.

SYMBOLS

a — constant in van der Waals equation
b — constant in van der Waals equation
B — second virial coefficient
c — velocity of propagation of pressure pulse in a liquid (8)
E — total energy per unit area of liquid surface (7), Young's modulus (9)
F — force; Helmholtz free energy (6), tension in a liquid (8)
$g(r)$ — radial distribution function
G — Gibbs free energy; shear modulus (9)
h — Planck constant
k — Boltzmann constant
K, K_s, K_T — bulk modulus, adiabatic and isothermal bulk modulus
l_f, l_v — specific latent heat of fusion, of vaporization
m — mass
N, N_A — number of molecules in assembly; Avogadro constant (number)
p — pressure
Q — quantity of heat; volume flow per second of liquid through a tube (9)
r, r_0 — intermolecular separation, normal equilibrium intermolecular separation
R — gas constant
S — entropy; rate of shear (10)
T, T_c, T_r — thermodynamic (absolute) temperature, critical temperature, triple point temperature

U	internal energy
v	velocity
V	volume
z	number of nearest neighbours (coordination number) in a solid (lattice) or a liquid (quasi-lattice)
α	polarizability
β	yield stress (10)
β_T	isothermal compressibility
γ	free surface energy, surface tension
δ	phase angle
ϵ	maximum (negative) value of intermolecular potential energy; shear strain (10)
η	viscosity
λ	elongational viscosity
ν	kinematic viscosity
μ	electric moment (4), Bingham viscosity (10)
ρ	density
$\rho(r)$, ρ_0	number density and mean number density of molecules
σ	molecular diameter
τ	shear stress
ϕ	mutual potential energy of two molecules
ω	angular velocity

UNITS

SI units are used throughout this book. These are explained, for example, in the Royal Society booklet *Quantities, Units and Symbols* (1975). A list of six of the base SI units, together with derived units, prefixes, conversion factors and numerical values for some important constants follows.

BASE SI UNITS

Quantity	*Name*	*Symbol*
length	metre	m
mass	kilogram	kg
time	second	s
electric current	ampere	A
thermodynamic temperature	kelvin	K
amount of substance	mole	mol

DERIVED SI UNITS

Quantity	*Name*	*Symbol*
force	newton	$N = kg\ m\ s^{-2}$
energy	joule	$J = kg\ m^2\ s^{-2}$
electric charge	coulomb	$C = A\ s$
electric potential difference	volt	$V = J\ A^{-1}\ s^{-1}$
electric intensity	volt per metre	$V\ m^{-1}$
frequency	hertz	$Hz = s^{-1}$
velocity	metre per second	$m\ s^{-1}$
acceleration	metre per second²	$m\ s^{-2}$
pressure	pascal (newton per square metre)	$Pa\ (N\ m^{-2})$

PREFIXES FOR SI UNITS

Multiple	*Prefix*	*Symbol*	*Multiple*	*Prefix*	*Symbol*
10^{-12}	pico	p	10^{3}	kilo	k
10^{-9}	nano	n	10^{6}	mega	M
10^{-6}	micro	μ	10^{9}	giga	G
10^{-3}	milli	m	10^{12}	tera	T

OTHER UNITS

Quantity	*Name*	*Symbol*
viscosity	poise ($g\ cm^{-1}\ s^{-1}$)	$P = 10^{-1}\ kg\ m^{-1}\ s^{-1}$
kinematic viscosity	stokes ($cm^2\ s^{-1}$)	$St = 10^{-4}\ m^2\ s^{-1}$

PHYSICAL CONSTANTS

h	Planck constant	$6{\cdot}62 \times 10^{-34}\ J\ s$
k	Boltzmann constant	$1{\cdot}38 \times 10^{-23}\ J\ K^{-1}$
N_A	Avogadro constant	$6{\cdot}02 \times 10^{23}\ mol^{-1}$
R	gas constant	$8{\cdot}314\ J\ K^{-1}\ mol^{-1}$

CONVERSION FACTORS

ångström	$1\ Å \equiv 10^{-10}\ m = 0{\cdot}1\ nm$
atmosphere	$1\ atm \equiv 101{\cdot}3\ kN\ m^{-2} = 1{\cdot}013 \times 10^{5}\ Pa$

A screw propeller acting in the 'fully cavitating' condition in the National Physical Laboratory ... of the National Physical Laboratory).

CONTENTS

CHAPTER 1
introduction

1.1. *The three phases of matter*

EVERYONE knows that matter can exist as a solid, a liquid or a gas and the more obvious differences between these three phases form a part of our day to day experience. A solid piece of matter, such as a metal, is one which possesses a certain definite shape which it retains under ordinary conditions. A liquid will take up the shape of the vessel into which it is poured and possesses a well-defined upper free surface. A gas must be contained in a closed container whose internal volume it fills completely. These are well-known facts and we must now look at these three phases from the molecular point of view. Indeed the modern approach is to start with the fundamental units—the atoms or molecules—of which a specimen of matter is composed and to estimate the forces which adjacent molecules exert on each other. From this ' microscopic ' approach an attempt is then made to predict the ' macroscopic ' or bulk properties of the specimen.

Let us then consider, in a simple way, the three phases of matter from the molecular point of view. In a solid the molecules are arranged in a regular spatial pattern known as a lattice and this gives the solid its compact, unchanging shape. In a liquid the individual molecules have more ' elbow room ' and more freedom to wander a little. In a gas each molecule can dash around all over the volume in which the gas is enclosed. A rough analogy to the three molecular arrangements is the following. Let us imagine a number of people. They can be sitting in the regularly arranged seats in a lecture theatre (solid) or they can be talking in small close groups with the odd person migrating here and there to another group (liquid) or they can be like soccer players moving around in a largely empty field (gas).

1.2. *Changes of phase*

Another familiar experience is that matter can change from one phase into another. A solid melts and changes into a liquid, which in turn boils and changes into a vapour (that is, a gas). The two temperatures at which these changes occur under ordinary conditions are the melting and boiling points. But we must be careful not to ascribe too much importance to these temperatures for the following reason. Water, for example, boils at 100°C because its vapour pressure is then equal to the pressure of the Earth's atmosphere. If the ambient pressure above the liquid changed the water would boil at a different temperature. So we see that the boiling point is a property of both the liquid and the surrounding ambient pressure and its rate of change with this pressure is high. For example, the reduction in atmospheric pressure at the top of a high mountain is enough to reduce the boiling point of water considerably. In the same way the melting point varies with the ambient pressure though much less markedly. These facts clearly suggest that we should isolate a specimen of matter from its surroundings and seek for two other temperatures of a more fundamental nature, in the sense that they are properties of the substance only, to replace the boiling and melting points.

So we consider a cylinder, with a movable piston, from which the air inside has all been previously removed. If the cylinder, whose walls are assumed to be transparent, contains a liquid and its vapour as in fig. 1.1 then the rate at which molecules leave the liquid surface to enter the space above (evaporation) is equal to the rate at which the vapour molecules return to the liquid (condensation). This dynamic two-way traffic is said to represent a state of thermodynamic equilibrium.

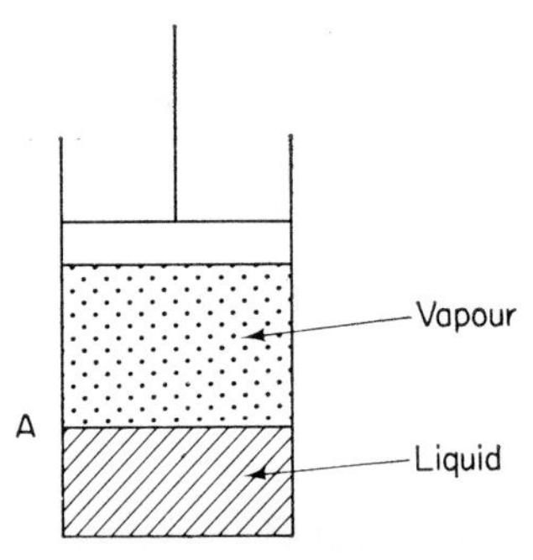

Fig. 1.1. A cylinder, with a movable piston, containing a liquid and its vapour in thermodynamic equilibrium.

Suppose that we start with water and water vapour at 15°C in our cylinder. The vapour and liquid regions will be clearly separated by the boundary surface A. The liquid density ρ_l will be just less than 1000 kg m^{-3}, the vapour density ρ_V about $1{\cdot}37 \times 10^{-2}$ kg m^{-3} and the vapour pressure p about 0·017 atm. Next suppose that the temperature is raised whilst the volume is kept constant by keeping the piston fixed. This causes the liquid to expand and so decrease in density; more molecules leave the liquid to enter the vapour phase causing ρ_V and p to increase. As this rise in temperature continues ρ_l decreases all the while and ρ_V increases until at a temperature of about 374°C these densities become equal; at the same time the dividing boundary A between the two phases disappears and there are no longer two different regions visible within the cylinder. As the temperature is raised still further this uniform nature of the contents of the cylinder persists. The temperature at which the liquid–vapour boundary disappears is called the critical temperature, T_c. It is the highest temperature at which the liquid phase can exist and is one of the two 'fundamental' temperatures we are seeking. At temperatures above T_c all the properties which distinguish a liquid from a gas disappear and we simply have 'fluid-like' properties. Furthermore changing the pressure by moving the piston does not change the situation and does not result in a separation into liquid and gaseous phases. For water the critical data are $T_c = 374{\cdot}15°C$, $p_c = 218{\cdot}3$ atm and $\rho_c = 320$ kg m^{-3}.

If the contents of our cylinder are now cooled at constant volume below T_c the original liquid–vapour mixture reappears and eventually, at a certain temperature T_r, ice begins to form. If the cooling process is arrested so that the temperature is kept constant at T_r the solid, liquid and vapour can coexist in thermodynamic equilibrium; so we call T_r the triple point since the three phases then coexist. If the temperature falls below T_r the liquid phase disappears at once and changes into ice, and we then have the solid and vapour phases in equilibrium. Thus T_r is the temperature below which the liquid phase cannot exist; it is the lower 'fundamental' temperature we are looking for. For water T_r is, by the definition of the kelvin, 273·16 K. This is about 0·01°C, just a little above the normal freezing point (273·15 K).

Experiments such as those just described were first carried out in the last century by de la Tour and later by Andrews.

1.3. *The phase diagram*

Using the apparatus shown in fig. 1.1 a (p,T) or (p,V) diagram can be drawn to illustrate the equilibrium between the three phases nside the cylinder. In fig. 1.2 the (p,T) diagram for a simple ubstance, like liquid argon, is shown. (The diagram is not drawn o scale but is meant to show the general features only.) It is called phase diagram.

RC is the vapour pressure curve and shows the (p,T) variation when the cylinder contains only liquid and vapour. It shows the way in which the saturation vapour pressure varies with temperature. OR is called the sublimation curve and arises when the cylinder ontains only solid and vapour. RA shows what happens when the ylinder contains solid and liquid only and is the melting curve. It is steep and almost vertical and illustrates how the melting point aries with pressure. In general the melting point rises as the pressure rises but for water it is lowered; thus for water the portion RA slopes in a direction opposite to that shown in fig. 1.2.

It should be noted that the point A is not the end of the curve RA and no such terminal point has been observed experimentally. On the other hand the curve RC can have only the extent shown since R, the triple point, and C, the critical point, represent the lower and upper limits of the existence of the liquid phase. The point R is the only one where the solid, liquid and gas regions meet. The shaded area in the diagram represents the region with which this book deals.

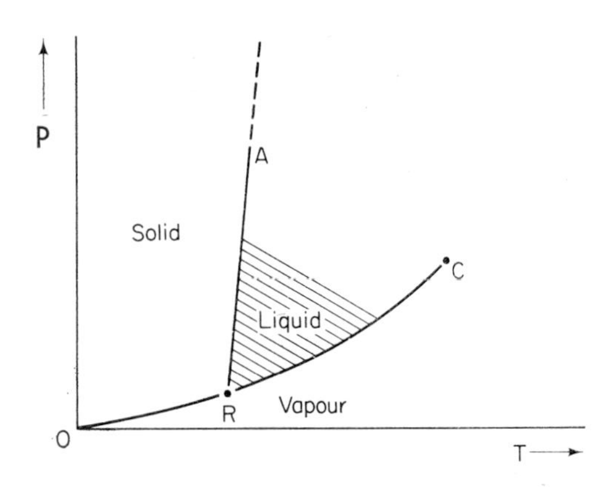

Fig. 1.2. The phase diagram of a simple substance.

This region is a very small part of the whole phase diagram. However it is of great practical importance for two reasons. Nearly all substances melt and boil at readily attainable temperatures. Secondly, it so happens that the freezing and boiling points of water are 0°C and 100°C while the ambient temperature on the earth's surface varies from roughly −60°C to +60°C. This means that the freezing and evaporation of water are of fundamental importance biologically and environmentally.

In table 1.1 we give the critical and triple points for a few substances. The last column shows that the ratio T_c/T_r ranges from 1·41 to 2·47. Indeed for most substances T_c/T_r and also V_c/V_r lie in the range 1 to 5. This shows just how small a region of the whole phase diagram is that corresponding to the liquid phase.

Substance	T_c/K	T_r/K	T_c/T_r
Ar	151	83·00	1·82
CO_2	304	216·00	1·41
H_2O	674	273·16	2·47

Table 1.1. Values of T_c, T_r and T_c/T_r for three substances.

1.4. *Other preliminary observations*

Liquids are a phase of matter intermediate between the solid and gaseous phases. In some ways a liquid resembles a solid, in other ways a gas. A solid and liquid are similar in the manner in which the molecules in both crowd together while the molecules in a gas move apart to fill the whole volume available; so we say that solids and liquids possess cohesion exerted by the cohesive forces between their molecules. A similarity between liquids and gases is that they cannot support shear stresses and give way by flowing when subjected to them; solids, however, can support these stresses. Again, consider the effects of temperature; at very low temperatures matter is solid, at intermediate ones liquid and at very high temperatures it is gaseous.

This intermediate role of the liquid has played a great part in our understanding of liquids and most theories of liquids are 'extrapolations' from gas-like or solid-like theories. The theory of gases was largely developed in the nineteenth century whilst the theory of solids was begun in the early twentieth century by workers such as Einstein, Nernst, Debye and Born. The history of the kinetic theory of liquids is one of a number of fundamental advances, often separated by quite long intervals; since the second world war considerable steady progress has been made.

Another point worth mentioning is this. Although we have been emphasizing the importance of the molecular or miscroscopic approach to a study of liquids it should not be forgotten that the concept of a liquid as a *continuous* medium is still most useful. Much of the hydrodynamics originally laid down by Newton, Bernoulli and others remains perfectly valid today and is based on treating a liquid as a continuous medium.

CHAPTER 2

liquids as modified solids: the radial distribution function

THE spatial arrangement of the molecules in a liquid is best described by an important function called the radial distribution function. Before discussing this we must discuss the lattice model of a solid and the so-called quasi-lattice model of a liquid.

2.1. *The lattice model of a solid*

As we would expect, the similarity between a solid and liquid is most evident at temperatures near the triple point where the transition from solid to liquid occurs. So we start by describing the molecular arrangement in a solid.

Most solids are crystalline and at the absolute zero their atoms or molecules are at rest at certain points or ' sites '. These sites form a definite regular geometrical pattern, known as a lattice, which extends throughout the whole of the crystal. For example, in a ' simple cubic ' lattice (fig. 2.1) the sites would be situated at the eight corners of tiny cubes, all of the same size, and this arrangement would extend inside the whole interior of the solid. Because this regular geometrical pattern or ' order ' exists throughout the whole solid, we say that the solid possesses long-range order. If we choose any internal molecule as some reference point then the position of any other molecule however far away, will bear some definite geometrical relation to the position of this initial reference molecule. For example, in a simple cubic lattice let us suppose that the elementary unit cube has a side of length a; then if we choose a given molecule and extend the distance from it along the cube side by any distance na (where n is whole number) there will be another lattice site at that point at n cube sides away.

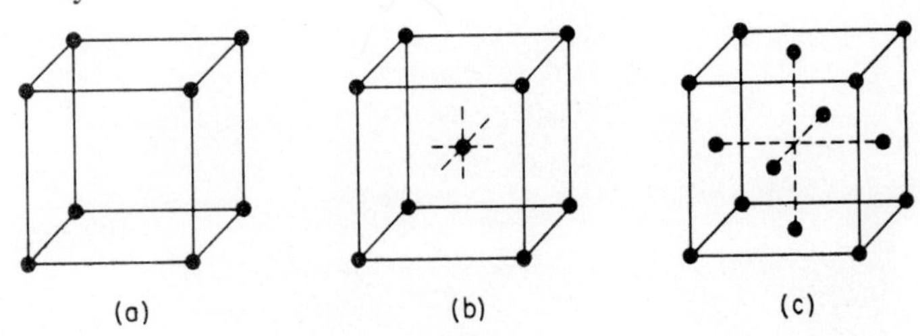

Fig. 2.1. Unit cells in (*a*) a simple cubic, (*b*) a body-centred cubic and (*c*) a face-centred cubic lattice.

Any typical molecule will also have a certain number z of nearest neighbours on its adjacent sites; z is called the coordination number of the lattice. If we choose some given molecule A, its z nearest neighbours are equidistant from it and so lie on a sphere of centre A. The next-nearest neighbours will lie on a second concentric sphere of larger radius and so on. These concentric spherical surfaces are often known as ' shells ' so that we talk about the nearest-neighbour shell and so forth.

Now consider what happens as our solid is heated up above the absolute zero. Each molecule starts to vibrate about its lattice point and the amplitude of each vibration increases as the temperature increases. Nevertheless the centres of vibration continue to remain at the well-defined system of lattice points so that the ' long-range order ' is still preserved.

We have already mentioned the simple cubic lattice; it has a coordination number of $z=6$. In the alkali metals we have a body-centred cubic lattice with one atom at a cube centre with another at each of the eight corners of the cube; here $z=8$. The inert gases, helium excepted, form a face-centred cubic lattice when the gases are solidified; here we have 8 atoms at the 8 cube corners and 6 others at the centre of the 6 cube faces. The distance between nearest neighbours is half the diagonal of a cube face and in this case $z=12$.

2.2. *The quasi-lattice model of a liquid*

In the previous section we discussed the lattice model of a solid and now consider what happens as the temperature is raised still further. A struggle then occurs within the lattice. On the one hand we have the 'ordering' influence of the intermolecular forces, which tries to preserve the geometry of the lattice; this is opposed by the increasing thermal vibrations of the molecules which tend to disrupt the lattice. Eventually the latter effect prevails and at that stage the solid melts into a liquid with the result that the original long-range order is destroyed but a good deal of short-range order remains. By this we mean that the molecular arrangement in the liquid near a chosen central molecule A will be a fairly regular one, very like that in the corresponding solid, but when we move out to distances of 30 Å = 3 nm or so from A then the positions there of the molecules will bear no spatial geometrical relations to the position of A. Thus there is local or short-range order just around any given molecule but no long-range order throughout the liquid. So we describe the molecular arrangement in a liquid as a quasi-lattice. We shall see how to deal with this situation in the next section.

When a solid melts into a liquid there is usually a volume increase of 5 to 15 per cent. This means that the molecules in a liquid have more ' elbow room ' than those in the solid. Most of our knowledge of these molecular arrangements comes from experiments using X-ray and neutron diffraction (see section 5.7). These results show that there is a break-down of the long-range order of the solid at the melting point but that short-range order still persists above this temperature when the substance is then in the liquid phase. One might also suppose that the volume increase on melting is due to a uniform increase in the distance between neighbouring molecules. On the contrary the diffraction experiments indicate that the increase in volume is due to a general decrease in the coordination number z. For example if z were 12 in the solid phase it might be reduced on average to 10 in the liquid phase; hence each molecule in the liquid would be surrounded on average by 10 nearest neighbours and two empty sites or ' holes '. So the greater volume of the liquid could be ascribed to a greater number of sites, most of which are occupied by a molecule and the remaining few by holes; for example, in the case just considered a fraction of 10/12 of the sites would contain a molecule but the remaining fraction would be holes. Put in another way we can suppose that a number of holes has been injected into the liquid to ' inflate ' its overall volume to about 10 per cent or so greater than that of the solid.

Bernal has picturesquely described a crystalline solid as a pile of molecules whereas a liquid is a heap. One can imagine a ' pile ' of bricks in which the bricks are laid regularly, layer upon layer, to form a three-dimensional rectangular stack; this is like the regular lattice of a solid. Again one can picture a number of bricks being tipped from a lorry in a somewhat pyramid-shaped ' heap '; the bricks will not be as closely packed as in the ' pile ' and this is more like the quasi-lattice of a liquid.

2.3. *The radial distribution function*

The lattice, with its regular geometry, provides the necessar description of the spatial correlations between molecules in crystalline solid. For a liquid the corresponding quantity is th *radial distribution function.* We shall confine our attention to simple liquid, such as liquid argon, in which the molecules ar spherically symmetrical. Each molecule can then be regarded as small sphere of diameter σ and the force between two molecules wi depend only on the separation of their centres.

Let us then take one molecule, of centre O, and consider th surrounding ' liquid ' environment as viewed by an observer at C At a distance r from O let the number of molecules per unit volum be denoted by $\rho(r)$, the number density. If we consider any straigl line passing through O then the value of $\rho(r)$ will vary with r; du to the nature of the short-range order $\rho(r)$ will be sometimes mor sometimes less than the mean number density ρ_0 throughout the whole liquid where

$$\rho_0 = \frac{\text{total number of molecules in the liquid}}{\text{total volume of the liquid}} \; .$$

This means that the ratio $\rho(r)/\rho_0 = g(r)$ can be sometimes greater or less than unity as r varies. This function $g(r)$ is the radial distribution function and the graph of $g(r)$ against r is of the general shape shown in fig. 2.2.

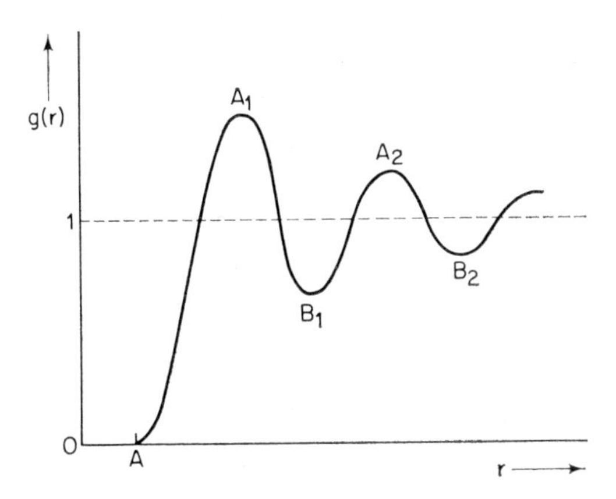

Fig. 2.2. The radial distribution function for a simple liquid.

For values of r between O and A, where OA is equal to the diameter of a molecule, the value of $g(r) = 0$; a little reflection shows that this must be so. For large r the value of $\rho(r)$ tends to ρ_0, that is, $g(r)$ tends o unity. But for distances of two or three molecular diameters the value of $g(r)$ departs appreciably from unity as shown by the points A_1, B_1, A_2, B_2, etc. This departure of $g(r)$ from unity is then a measure of the short-range order in the liquid around the chosen molecule O. The first peak A_1 corresponds to the inner shell of O's nearest neighbours, A_2 to O's second shell of neighbours and so on. The troughs B_1, B_2, etc., correspond to the ' emptier ' regions between hese shells. For a typical liquid, such as liquid argon at 84 K, the pproximate values of r at the first and second peaks are 4·0 and 7·0 Å espectively.

It is important to remember that $g(r)$ for any value of r represents he mean value, averaged over time, at that distance. Large statistical uctuations from this mean value will occur due to the thermal motion f the molecules. $g(r)$ is important in two respects. First, Debye howed that the diffraction pattern of a liquid is related to $g(r)$; this means that $g(r)$ can be obtained from experiments involving the diffraction of X-rays and neutrons by liquids. Secondly, if it is assumed that the total potential energy of the liquid is the sum of contributions between pairs of molecules, then quantities like the pressure and the compressibility can be calculated directly from $g(r)$. (We might note, in passing, that in an ideal gas, where there are no correlations, $g(r)=1$ everywhere, that is, for all values of r.)

Some experimental radial distribution curves obtained by Eisenstein and Gingrich for liquid argon using X-ray diffraction are shown in fig. 2.3. The values of $g(r)=\rho(r)/\rho_0$ are plotted as ordinates and the values of r/σ as abscissae, where $\sigma=3{\cdot}42$ Å is the diameter of an argon molecule. To simplify matters the ordinates of all the curves, except curve 1, have been displaced vertically.

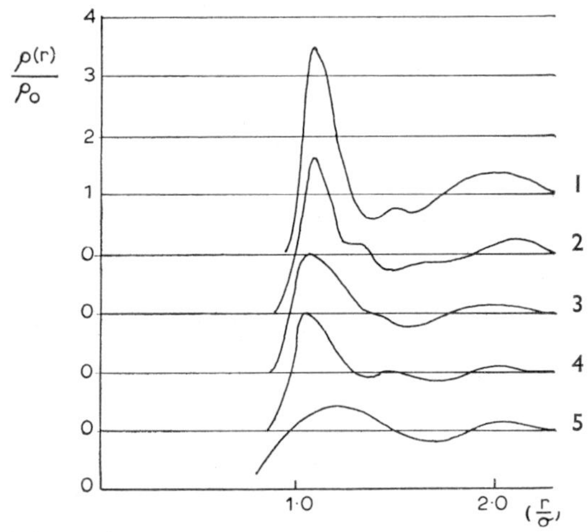

Fig. 2.3. Radial distribution function for liquid argon at different temperatures: (1) 84·4 K, (2) 91·8 K, (3) 126·7 K, (4) 144·1 K and (5) 149·3 K (This curve is reproduced from *The Liquid State* by J. A. Pryde.)

Curve 1 is for the triple point (84·4 K) and shows a good deal o short-range order. Curves 2, 3 and 4 correspond to successive highe temperatures and illustrate how this order gradually disappears as th temperature rises. Finally curve 5 shows the results for a temperatur very near the critical point where the gaseous phase is bein approached.

References

(1) Pryde, J. A., *The Liquid State*, Chapter 3 (Hutchinson University Library London, 1966).

(2) Tabor, D., *Gases, Liquids and Solids*, Chapter 11 (Penguin Library c Physical Sciences, 1969).

CHAPTER 3

the 'gas-like' approach to a liquid and the van der Waals equation

JUST as the similarity between a liquid and solid is more evident near the triple point so is that between a liquid and gas more evident near the critical point. This chapter deals with the gas-like approach to liquids.

3.1. *The ideal or perfect gas*

We start by considering the simplest conceivable sort of gas, the ideal or perfect gas. This consists of identical molecules, each a negligibly small point mass, moving around at random in a closed container. It is assumed that there are no intermolecular forces between any two molecules and this means that there is no mutual potential energy between them either. This further means that the total internal energy U of the gas is simply the sum of the kinetic energies of the individual molecules and the result turns out to be

$$U=\frac{3}{2}NkT \tag{3.1}$$

for a gas of N molecules at thermodynamic temperature T; the constant k is the Boltzmann constant.

For one mole of gas we usually write $U=(3/2)RT$ where $R=N_{A}k$ and N_{A} is equal to the number of molecules which one mole of a gas contains. N_{A} is the Avogadro constant and has the value of $6{\cdot}02\times10^{23}\ \mathrm{mol^{-1}}$; R is the gas constant and its value is

$$8{\cdot}314\ \mathrm{J\ K^{-1}\ mol^{-1}}.$$

A simple form of the kinetic theory of gases then leads to the well-known equation of state

$$pV=RT \tag{3.2}$$

for one mole of gas (the form we shall use throughout since for any quantity n moles it becomes simply $pV=nRT$). If the temperature is kept a constant this result reduces to Boyle's law, namely,

$$pV=\text{constant} \tag{3.3}$$

and the corresponding graph of p against V at this constant temperature is called the (p,V) isotherm.

3.2. *How van der Waals modified the ideal gas equation*

A real gas is, alas, not as simple to deal with as the ideal gas of the previous section. Still, if the density of a real gas is *low*, so that its molecules are far apart for most of the time, we can then assume that there are no effective intermolecular forces present. So at such low densities we would expect equations (3.2) and (3.3) to be obeyed by a real gas and this is found to be so. But at higher densities, the molecules of our real gas will be crowded closer together and we can no longer ignore the forces between its molecules.

What can we say about the forces between two such real molecules? The fact that there are cohesive forces holding the molecules together in solids or liquids suggests that forces between molecules are attractive (see section 1.4). Yet it is extremely difficult to compress a solid or liquid and so we conclude that, below a certain intermolecular separation, the force between two real molecules becomes strongly repulsive. So for two real molecules we must take account of *repulsions* when they are extremely closely together and of *attractions* when they are somewhat further apart. About a century ago van der Waals showed how this could be done.

To deal with the repulsions the two molecules must be regarded not as mass points but rather as rigid spheres having a finite diameter σ. The minimum possible distance between the centres of the two molecules will then be σ (see fig. 3.1).

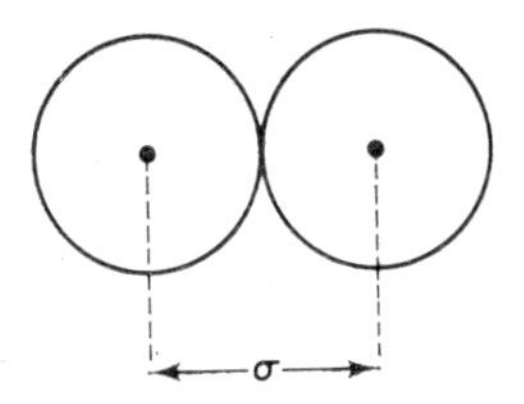

Fig. 3.1. Two 'rigid' spherical molecules, each of diameter σ, with their centres at the minimum possible distance of σ apart.

This idea of replacing 'point' molecules by rigid spheres of finite diameter σ is a convenient way of saying that when the distance r between two molecules decreases to σ a very strong or infinite repulsion between them occurs. Since the spheres are rigid it will just not be possible for the two molecular centres to get closer together than σ. If V is the volume of the enclosure occupied by the gas then the total effective volume available for the molecules to roam in is not V as for point molecules but something less, say $(V-b)$, because of the finite size of the molecules. In fact b turns out to be four times the total volume of all the molecules of the gas; that is $b=2N\pi\sigma^3/3$ if there are a total of N molecules.

Now consider the attractions. When the distance r between two molecules exceeds σ the force between them is an attraction; when r becomes fairly large, say about five molecular diameters or even less, this attraction is virtually zero. Van der Waals dealt with all these attractions by assuming that they were equivalent to some internal pressure p' causing the molecules to crowd together; this means that the applied pressure p of the ideal gas case must be replaced by $p+p'$. Van der Waals further argued that p' is proportional to the square of the density of the gas; this means $p' \propto 1/V^2$ for a constant mass of gas. Thus (for one mole) the perfect gas equation (3.2) becomes modified to

$$\left(p+\frac{a}{V^2}\right)(V-b)=RT \tag{3.4}$$

which is van der Waals' famous equation.

If we give T some constant value we can then plot the (p,V) isotherm for that temperature; this is then repeated for other fixed values of T. In this way we can obtain the family of van der Waals isotherms shown in fig. 3.2.

The lowest isotherm for $T=T_1$ has a minimum at A_1 and a maximum at B_1. Along the portion A_1B_1, dp/dV is positive, that is, the volume increases as the pressure is increased; clearly this cannot

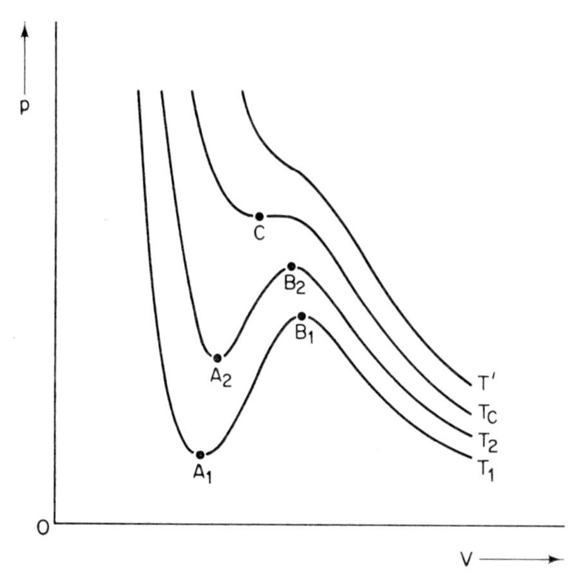

Fig. 3.2. Family of van der Waals isotherms obtained from a mathematical plot of equation (3.4) for various values of T.

happen for a real gas so that the region A_1B_1 does not 'make sense' physically. An isotherm at a higher temperature $T=T_2$ has the same general sinuous shape except that the minimum A_2 and maximum B_2 have moved closer together. As T rises above T_2 the corresponding turning points move still closer together until at some temperature $T=T_c$ they merge into a single point C which is a point of inflexion. For temperatures $T'>T_c$, p decreases continuously as V increases. The above discussion is what a mere mathematical plot of equation (3.4) tells us. How does this compare with what occurs in practice? We answer this question in the next section.

3.3. *Comparison with experiment*

We return to the cylinder and piston experiment (section 1.2) using water as the enclosed substance. Let us start at 15°C ($T_1 \simeq 288$ K) and suppose that the piston is drawn well out so that the cylinder contains only water vapour; the point D_1 in fig. 3.3 will then represent this initial state. If the piston is next moved in, while the temperature T_1 is kept constant, the isotherm follows the path D_1E_1. At E_1 the value of p is 0·17 atm and at this stage drops of water start to condense on the inside walls and the piston base. As the volume is further reduced beyond that at E_1 more and more water condenses but the pressure inside remains a constant; thus the curve to the left of E_1 follows a horizontal, constant-pressure path until we reach the point F_1 which corresponds to the case where *all* the vapour has condensed and the piston is now resting on the face of the liquid itself. The horizontal portion E_1F_1 of the curve is therefore a two-phase region of liquid–vapour equilibrium and the constant pressure corresponding to E_1F_1 is the saturation vapour pressure at the temperature T_1. If the piston is now pushed in further we shall be compressing the liquid; as already mentioned this is not easy and requires a huge change in pressure to cause even a small volume change. This is why the portion F_1G_1 of the curve rises almost vertically.

Let us now repeat the experiment at a higher temperature, say 100°C ($T_2 \simeq 373$ K). Starting at D_2 the curve follows the path D_2E_2 to E_2 where condensation sets in; at E_2, $p=1$ atm. By reducing the volume continuously as before the remaining part $E_2F_2G_2$ of the curve is traced out. The straight portion E_2F_2 is shorter than E_1F_1 the volume at E_2 is less than that at E_1 while that at F_2 is greater than that at F_1. If the experiment is repeated for temperatures higher than T_2 the ends of the horizontal portions such as E_2F_2 move closer together until at a certain temperature T_c they merge into the same point C. This is the critical point which we earlier called a point of inflexion.

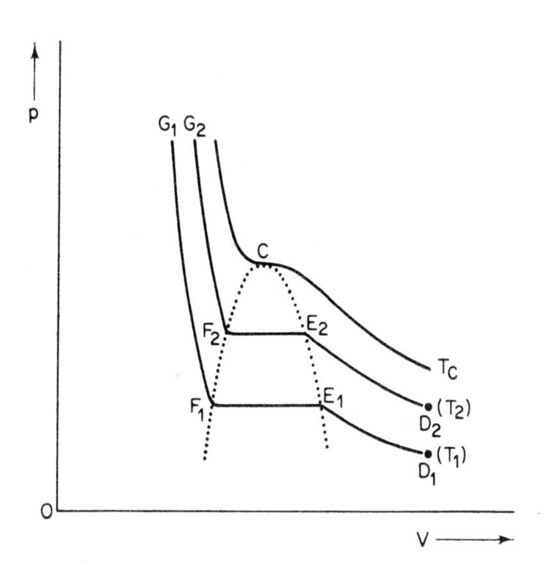

Fig. 3.3. (p,V) isotherms of a substance obtained from experiment.

3.4. *Reconciling the experimental and the van der Waals isotherms*

To reconcile the experimental curves of fig. 3.3 with the van der Waals isotherms of fig. 3.2 we consider one typical van der Waals isotherm, as in fig. 3.4, corresponding to $T = T_1$.

We then draw the horizontal line F_1E_1 so that the two shaded areas are equal. This is called the " rule of equal areas " and it ensures

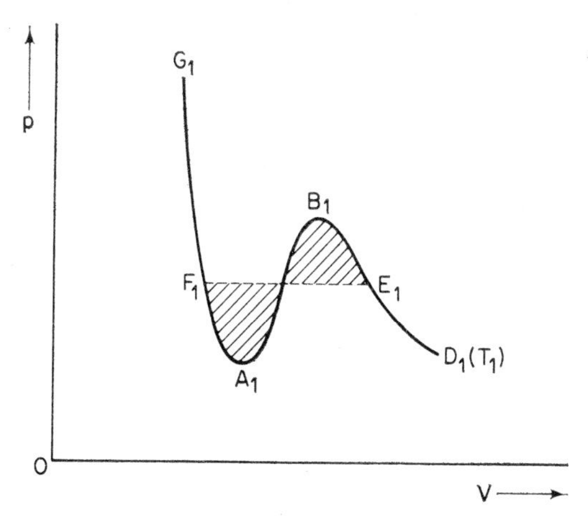

Fig. 3.4. Reconciling the experimental and the van der Waals isotherms.

that the position of F_1E_1 is that corresponding to the liquid–vapour equilibrium. The reason for this is a thermodynamic one. For equilibrium a certain thermodynamic function—the Gibbs function, G—must have an equal value for unit mass of the liquid and for unit mass of the vapour in equilibrium with it (see Appendix (1)). It is not difficult to show that this demands that the two areas shaded should be equal.

The similarity between the curve of fig. 3.4 and the experimental isotherm for $T = T_1$ in fig. 3.3 is then obvious. In both cases the portions D_1E_1 represent the vapour phase, the horizontal portions E_1F_1 the liquid–vapour two-phase equilibrium and the portions F_1G_1 the liquid phase. The portion A_1B_1 of fig. 3.4, where $\mathrm{d}p/\mathrm{d}V$ is positive, is not physically realizable, as already mentioned in section 3.2. So we are left with the regions F_1A_1 and E_1B_1 where $\mathrm{d}p/\mathrm{d}V$ is still negative and we enquire whether they can be realised in practice. The answer is 'yes' *under certain conditions* and for this reason F_1A_1 and E_1B_1 represent *metastable* phases.

Consider first the region F_1A_1. As we move down the region G_1F_1 of the curve we would normally expect the liquid to split up into vapour and liquid at the point F_1 and follow the path F_1E_1. But if the liquid is in a vessel with smooth walls and is free of dust particles or similar 'nuclei' then decreasing the pressure on the liquid to a point below that corresponding to F_1 may enable at least a part of the region F_1A_1 to be realized in practice. This is because there are no nuclei present on which vapour bubbles can form. Along F_1A_1 the liquid is said to be *superheated.* All this is the basis of the bubble chamber and also explains how a liquid can withstand tension (see Chapter 8).

Consider, secondly, the region E_1B_1. As we move along the curve from D_1 to E_1 we expect condensation to set in at E_1. If, however, there are no suitable 'condensation' nuclei present then the portion E_1B_1 of the curve can be realised, at least in part. Along E_1B_1 the vapour is said to be *supersaturated.* All this is used in the Wilson Cloud Chamber where ionizing particles act as condensation nuclei.

If the temperature is raised above T_1 the points such as F_1, A_1, B_1 and E_1 on the corresponding van der Waals isotherms all move closer together; when $T = T_c$ these four points merge into the single point C on the T_c isotherm of fig. 3.2 which we identify with the critical experimental isotherm of fig. 3.3.

3.5. *The critical point*

In fig. 3.2 the tangent to the critical isotherm $T = T_c$ at the point C is horizontal and C is also a point of inflexion. These two conditions together with the fact that the point C lies on the T_c isotherm give us three equations which are satisfied by the quantities p_c, V_c and T_c

corresponding to C. Solving these three equations is perfectly straightforward and the results only will be given. They are

$$p_c = \frac{a}{27b^2}, \qquad V_c = 3b, \qquad T_c = \frac{8a}{27bR}. \tag{3.5}$$

The quantities a and b can then be found by solving two of these equations, usually the first and third; from this calculated value of b one can then compare the predicted value, $V_c = 3b$, of the critical volume with the experimental value. Typical experimental values are $V_c = 2{\cdot}34b$ for argon and $V_c = 2{\cdot}44b$ for hydrogen both of which are low compared with the predicted value.

Again equations (3.5) predict a value of $8/3 = 2{\cdot}67$ for the dimensionless quantity RT_c/p_cV_c; the experimental values for argon and hydrogen are 3·43 and 3·28 respectively. Most of the experimental values of RT_c/p_cV_c are around 3·5.

These considerations show that van der Waals' equation is not very well obeyed near the critical point. We must not be too surprised about this since at the high densities in the critical region the molecules will be pretty closely packed. This means that the effects of intermolecular forces will be considerable and the somewhat simple way in which van der Waals dealt with these forces will not be valid at these high densities.

3.6. *The virial equation of state*

Equation (3.2) describes the behaviour of a perfect gas but most real or imperfect gases do not obey this equation closely even at room temperatures. One way of dealing with an imperfect gas is to modify equation (3.2) and rewrite it in the form of equation (3.4), the van der Waals equation. Another way of describing a gas whose behaviour departs only a little from ideality is to replace the perfect gas equation $pV/RT = 1$ by the equation

$$\frac{pV}{RT} = 1 + \frac{B}{V} + \frac{C}{V^2} + \frac{D}{V^3} + \ldots \tag{3.6}$$

This is called a virial equation of state and describes the behaviour of an imperfect gas at fairly low densities. B is called the second virial coefficient, C the third and so on. The first virial coefficient is simply unity so that for a perfect gas all the virial coefficients, apart from the first, are zero. Thus the coefficients B, C, etc., are a measure of how much the behaviour of an imperfect gas departs from that of a perfect gas.

The second virial coefficient B is also of great importance because it can be related to the mutual potential energy of two of the gas molecules. This point is discussed further in section 4.5.

3.7. *Conclusion*

In this chapter we have dealt mainly with van der Waals' equation and we have seen that it can describe the gaseous phase, the liquid phase and liquid–gas equilibrium. In this ' gas-like ' approach to liquids what we have done essentially is to regard a liquid as a highly compressed gas. Indeed, in the critical region, the density can be increased continuously from a low value appropriate to a gas to a high value appropriate to a liquid.

CHAPTER 4

intermolecular forces in liquids

4.1. *The general nature of intermolecular forces*

We have already said something about the forces between two similar molecules in section 3.2. We saw there that at very small values of the separation, r, of the molecules the force, F, between them was highly repulsive. At larger r, the force F was attractive. Thus for some intermediate separation $r = r_0$, F must be zero where the repulsion balances the attraction. Again for very large r, where the molecules are too far apart to affect each other, F is again zero. In all this we emphasize that we are assuming that we are dealing with simple spherical molecules and that F depends on r only. We are also assuming that we have just the two molecules only with no other molecules near them.

We now use the above reasoning to try to sketch the (F,r) curve between two similar molecules. Before doing this we decide on the following sign convention: a repulsion between the molecules is shown as positive and an attraction as negative. The general shape of the (F,r) curve is then shown in fig. 4.1.

At the point A the value of F reaches its *maximum attraction* and then 'tails off' to zero as r increases. At $r = r_0$ we have the normal

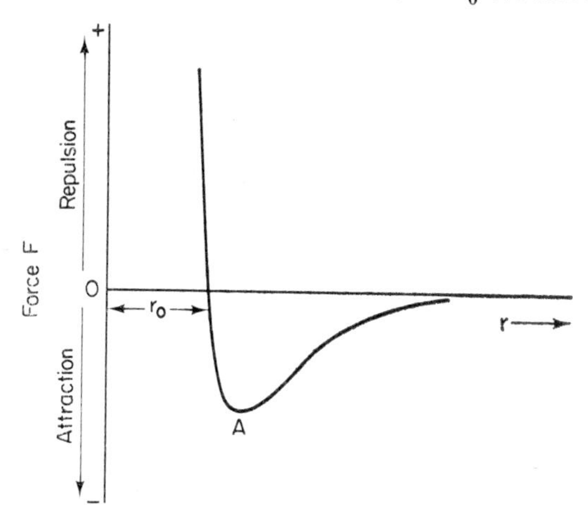

Fig. 4.1. Curve showing how the force, F, between two molecules varies with their separation, r.

equilibrium separation of the two molecules; a slight increase or decrease in r causes an attraction or repulsion respectively.

4.2. *The mutual potential energy of two molecules*

We next consider the mutual potential energy ϕ of the two molecules. The potential energy $\phi(r)$ of two molecules at a distance r apart is defined as the work required to bring one molecule from infinity up to a distance r from the other one. In considering these concepts in molecular physics the ' energy ' approach is a more powerful tool than the ' force ' approach and we now consider the relation between ϕ and F. To increase the distance r between two molecules to $r+\mathrm{d}r$ the work necessary is $F\mathrm{d}r$; in increasing the separation the potential energy decreases from ϕ to $\phi-\mathrm{d}\phi$. Thus $F\mathrm{d}r=-\mathrm{d}\phi$, or

$$F = -\frac{\mathrm{d}\phi(r)}{\mathrm{d}r}. \tag{4.1}$$

This very important equation then gives us the relation between the (F,r) and the (ϕ,r) curves. The general shape of the (ϕ,r) curve is shown in fig. 4.2 and we now discuss its main features. We first note that the (ϕ,r) curve is of the same general shape as the (F,r) curve of fig. 4.1. At $r=r_0$, $F=0$ (see fig. 4.1) and so $\mathrm{d}\phi/\mathrm{d}r=0$; thus at $r=r_0$ the curve of fig. 4.2 has a minimum where ϕ has a value of $-\epsilon$, say. Left to themselves the two molecules will tend to remain at this separation r_0 which corresponds to the minimum of the potential energy. To separate the two molecules to a large distance apart r would have to be increased from $r=r_0$ to $r=\infty$ and the work that

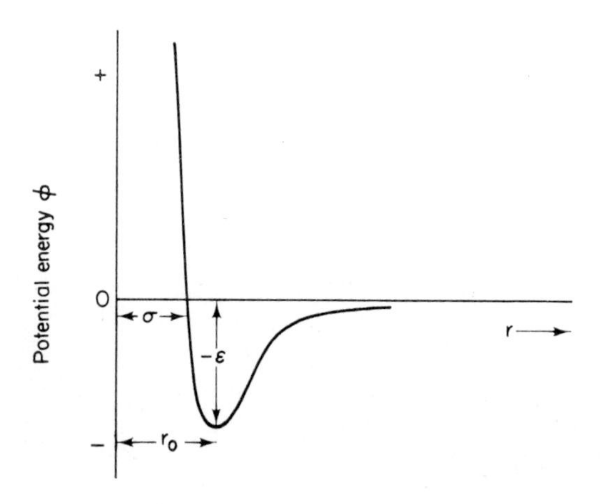

Fig. 4.2. The variation, with separation r, of the mutal potential energy $\phi(r)$ of a pair of molecules.

would have to be done is numerically equal to ϵ as we move from the minimum of the curve to its 'tail' as it approaches the r-axis at $r=\infty$. For this reason ϵ is often called the 'dissociation' or 'binding' energy.

The value of ϕ is zero at $r=\sigma$ and we now seek for a physical interpretation of σ. To do this we consider the two molecules to be initially at an infinite distance apart; this means that the value of ϕ will be zero initially. We next suppose that they start to move towards each other along AB, their line of centres, with initial negligibly small speeds (see fig. 4.3). So their initial total energy W, which is the sum of their initial potential energy ϕ and kinetic energy K, is virtually zero, that is, $W=\phi+K=0$. As the two molecules approach each other along AB both ϕ and K will vary but their sum W remains unchanged at zero. Eventually the two molecules collide and at this instant of collision the molecules are momentarily at rest so that their kinetic energy is zero, that is, $K=0$. So $\phi=0$ also at the instant of collision since the condition $W=\phi+K=0$ still holds. We also know that at collision the distance between the centres of the two spherical molecules is equal to their diameter (see fig. 3.1). Thus the value, σ, of r when $\phi=0$ in the curve of fig. 4.2. is the effective diameter of each molecule. In fact the equilibrium separation r_0 in fig. 4.2 is not much greater than σ and so r_0 is also a reasonable measure of the molecular liameter (see section 4.5).

Fig. 4.3. Two identical spherical molecules moving towards each other along their line of centres AB.

The value of r_0 can be estimated to a first approximation if the nolar volume of the liquid is known. This enables one to estimate he average volume per molecule. If this volume is pictured as eing a cube then the side of this cube is a good measure of the normal quilibrium separation, r_0, of the molecules. The details of this imple calculation are given in Chapter 3 of *Properties of Matter* by lowers and Mendoza. r_0 is about 2 to 5 Å in most cases (see also ble 4.1).

Our next task must be to try to explain the origin of the forces and nergies between molecules. We shall concentrate on the energy ϕ ther than on the force. Molecular physics shows that the value ϕ at any value of r is really the algebraic sum of two energies, an tractive energy ϕ_A and a repulsive energy ϕ_R. We now consider ese two energies in turn.

4.3. *The attractive energy*, ϕ_A

There are four types of attraction between molecules known as ionic, covalent, metallic and van der Waals binding. In liquids at ordinary room temperatures the relevant type of attraction is the last of these. These liquids can also exist as imperfect gases in the vapour phase and since the attractive forces and energies responsible for this ' imperfection ' were first studied by van der Waals his name has been kept to describe this type of binding.

To make progress we must note that there are two types of molecule, *polar* and *non-polar*, and our liquid will consist of one or other type. In general a molecule is built up of two or more atoms. The resulting molecule is electrically neutral and can be considered to consist of n positively charged protons each with a charge $+e$ and n negatively charged electrons, each with a charge $-e$. Viewed from the outside these charges can be considered as being equivalent to two single resultant point charges, a positive one of $+q=+ne$ at some point A and a negative one of $-q=-ne$ at a point B where AB is a small distance x (fig. 4.4). These two charges constitute an *electric dipole* of moment $\mu=qx$. This type of molecule in which the points A and B are separated by a finite distance x, usually of the order of 10^{-8} cm, is known as a polar molecule and is said to possess a permanent electric dipole moment. An example is the HCl molecule.

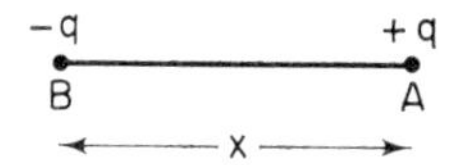

Fig. 4.4. A polar molecule pictured as an electric dipole.

In other types of molecule the two points A and B are coinciden and so the molecule has no permanent dipole moment and is non-polar Obvious cases giving non-polar liquids are the inert gases; these giv monatomic liquids and each atom (that is, molecule) has a spherically symmetric electron distribution around the nucleus. This is wh liquid argon is studied so widely as a typical non-polar liquid.

When a polar molecule is placed in an electric field the positiv and negative charges at A and B in the molecular dipole will be pulle apart even further and the increase in their separation will cause a increase μ_i in the dipole moment; μ_i is called an induced dipol When a non-polar molecule is placed in such a field the two point A and B will be actually separated from their point of coincidence by finite distance x' so that a dipole of moment $\mu_i=qx'$ is actually formed again we refer to μ_i as an induced dipole. The external field actin on our typical molecule can be produced either by the permaner dipole of a neighbouring polar molecule or, for example, by the field between the plates of a charged capacitor.

So, to sum up, when we have two polar molecules of a liquid near each other the attractive energy ϕ_A between them can be made up of three factors thus:

(*a*) Their permanent dipoles can interact giving them a mutual potential energy of ϕ_S.

(*b*) The permanent dipole of each can interact with the induced dipole in the other; this gives a mutual potential energy of ϕ_i.

(*c*) There is a third contribution, not immediately apparent, which is called the dispersion or London energy ϕ_L whose origin we shall discuss more fully below.

It is assumed that these three contributions can be considered separately and then added to give the total energy; thus

$$\phi_A = \phi_S + \phi_i + \phi_L. \tag{4.2}$$

In more advanced books (see, for example, *The Liquid State*, Chapter 4, by J. A. Pryde) it is shown that each of the three con-ributions on the right-hand side of equation (4.2) is proportional to $/r^6$ where r is the distance between the two molecules. From now n, however, we shall consider a simple non-polar liquid like argon o that the only contribution to the van der Waals energy ϕ_A will e ϕ_L since ϕ_S and ϕ_i will not arise for two molecules not possessing permanent dipole.

What, then, is the energy ϕ_L whose existence at first seems a uzzle? On the face of it one would not expect two non-polar molecules (1) and (2) to attract each other. Yet they clearly do ecause of the cohesion of a liquid like liquid argon. In the above iscussion it was stated that a non-polar molecule has no electric ipole because the electron shells are symmetrical with respect to the ucleus. But this statement is only valid *as a time average*. The ectrons are moving rapidly around the nucleus and the ‘ centre of avity ’, B, of their charges is not always coincident with that of the ositive nuclear charges, A. So at any given instant even a non-polar olecule, such as (1) in fig. 4.5, is a transitory dipole of moment μ.

g. 4.5. Showing how the transitory dipole μ in the non-polar molecule (1) can produce an induced dipole μ' in a second non-polar molecule (2).

At a point at a distance r along the axis of this dipole (1) the electric field E produced is proportional to μ/r^3 (a result readily found in books on electricity). So we can write

$$E = \frac{\beta\mu}{r^3}$$

where β is a constant.

If a second identical non-polar molecule (2) is placed at this point it will be polarized by the field of molecule (1) and the dipole μ' induced in it will be proportional to the strength of the field. So

$$\mu' = \alpha E$$

where α is known as the polarizability of the molecule.

Again from electrostatics we know that the potential energy of a dipole μ' in a field E is $\phi = -\mu' E$ which we can write in this case as

$$\phi_A = \phi_L = -\alpha E^2$$

$$= -\frac{\alpha\beta^2\mu^2}{r^6} \quad (4.3)$$

since the London energy is the only contribution to ϕ_A in this case This shows that the mutual potential energy of ϕ_L between the twc dipoles of fig. 4.5 is proportional to $1/r^6$. The treatment given abov is much over-simplified. For a full treatment the methods o quantum mechanics are needed. This was first given by F. Londo in 1930. He pictured the two neighbouring molecules as tw rapidly fluctuating dipoles, that is, ones whose dipole moments ar varying rapidly. London's result showed that $\phi_L \propto 1/r^6$ as in th simple treatment above. Another important point is that thes London forces *always* exist between two molecules and are alway *attractive.* Indeed this comes out in the simplified treatment give above. The two dipoles μ and μ' in fig. 4.5 attract each other in th configuration shown; if however the charges in dipole (1) are inter changed, so will those in (2) be interchanged and there will agai be an attraction between the dipoles. (We have a somewhat simila case in magnetism when we have the attraction between a magnet an a soft iron cylinder; reversing the magnet end-to-end also revers the induced moment in the soft iron and so the iron cylinder always attracted to the magnet.)

As a result of the above discussion we can write the attractiv van der Waals energy between our two non-polar molecules as

$$\phi_A = \frac{-A}{r^6} \quad (4.$$

where A is positive. The corresponding force between the molecules is

$$F_A = \frac{-d\phi_A}{dr} = -\frac{\text{constant}}{r^7}$$

where the constant is positive; F_A is known as the van der Waals force of attraction. These van der Waals forces are much weaker than the ionic and covalent forces existing in solids.

4.4. *The repulsive energy ϕ_R*

When our two molecules get very close together a very strong repulsion occurs between them. This repulsion is due to two effects. First, as the two molecules get nearer to each other their electron clouds eventually penetrate one another and this causes a strong repulsion between the two nuclear charges; this is because these two charges are no longer shielded electrostatically from each other by these clouds. The second effect occurs as a result of the Pauli exclusion principle which implies that two electrons of the same energy cannot be in the same region of space. So bringing two identical molecules together means 'packing' more electrons into the same element of space than is allowed by the Pauli principle. This causes a distortion of the electron clouds which manifests itself as a strong repulsion.

These considerations lead to the conclusion that the repulsive energy between two molecules is of the form

$$\phi_R = \frac{B}{r^n} \tag{4.5}$$

where B and n are positive. n is about 12 for the non-polar liquids we are discussing.

4.5. *The Lennard-Jones potential energy*

As already mentioned in section 4.2, the resultant intermolecular potential energy is the algebraic sum of ϕ_A and ϕ_R, so that from equations (4.4) and (4.5)

$$\phi = \frac{B}{r^n} - \frac{A}{r^6}.$$

Lennard-Jones and Devonshire in 1937 suggested an empirical form of this energy with $n = 12$ which they found to be suitable for the energy between two non-polar molecules and also for some weakly polar molecules. Using this value of n we then have

$$\phi = \frac{B}{r^{12}} - \frac{A}{r^6} \tag{4.6}$$

which is often called the 12–6 function. The graph of this 12–6 function has the same general shape as that shown in fig. 4.2 and our next step will be to relate the quantities σ, r_0 and ϵ in this graph to the quantities A and B in equation (4.6). Considering simultaneously the graph of fig. 4.2 and equation (4.6) we can make the following observations. When $r=\sigma$, $\phi=0$; so on substituting in equation (4.6) this gives

$$\frac{B}{A}=\sigma^6. \qquad (4.7)$$

Again when $r=r_0$,

$$\frac{\mathrm{d}\phi}{\mathrm{d}r}=0$$

which gives

$$\frac{A}{2B}=\frac{1}{r_0{}^6}. \qquad (4.8)$$

Finally when $r=r_0$, $\phi=-\epsilon$; this, together with (4.8) gives us

$$\frac{A^2}{4B}=\epsilon. \qquad (4.9)$$

So from equations (4.7) and (4.9) we have $A=4\epsilon\sigma^6$ and $B=4\epsilon\sigma^{12}$ and substituting these vales for A and B in (4.6) gives

$$\phi=4\epsilon\left[\left(\frac{\sigma}{r}\right)^{12}-\left(\frac{\sigma}{r}\right)^{6}\right]. \qquad (4.10)$$

Again from equations (4.7) and (4.8) we get

$$r_0{}^6=2\sigma^6$$

or

$$r_0=1{\cdot}12\sigma \qquad (4.11)$$

which shows that r_0 is about the same distance as σ. In other word both σ and r_0 can be taken as good measures of the diameter of th molecule.

Two molecules which are nearest neighbours in a liquid wi normally be at a distance $r=r_0$ apart, the normal equilibriur separation. So if two molecules are at *twice* this distance apart th mutual potential energy between them is the value of ϕ obtained b substituting $r=2r_0$ in (4.10). If we do this we get $\phi=-\epsilon/32$. S if the separation of two molecules is doubled from the equilibriur separation r_0 to $2r_0$ their mutual potential energy decreases from $-\epsilon$ t about $-\epsilon/30$; in other words the (ϕ,r) curve of fig. 4.2 quickly tenc to the r-axis when $r>r_0$. So whilst two molecules which are neare neighbours in a liquid are strongly bound together by an energy two molecules which are *next* nearest neighbours are bound togeth much more loosely. Thus for most purposes we can ignore bonds other than those between nearest neighbours. To cut each nearest-neighbour bond requires an energy ϵ.

Finally we consider briefly how the values of σ and ϵ for argon, say, can be determined in practice. There are two ways of doing this. One is by considering the properties of argon when it exists as an imperfect gas and relating the second virial coefficient B calculated on the basis of equation (4.10) to the value obtained experimentally. The detailed theory of imperfect gases shows that B can be expressed as

$$B = -2\pi N \int_0^\infty [\exp(-\phi(r)/kT) - 1] r^2 \mathrm{d}r \tag{4.12}$$

for a gas of N molecules. The *theoretical* value of B can be obtained by substituting the $\phi(r)$ of equation (4.10) into equation (4.12). Comparison between this theoretical value of B and its experimental value then yields the values of σ and ϵ. The second method is to consider argon in its crystalline solid phase at absolute zero; the total internal energy of this crystal lattice can be calculated in terms of ϵ and σ and can then be related to the zero-point latent heat of sublimation.

In table 4.1 the values of σ and $T_0 = \epsilon/k$ for a few simple molecules are listed. The reason why we tabulate $T_0 = \epsilon/k$ rather than ϵ is that the mean kinetic energy of a molecule at a temperature T is of the order of kT (see equation (3.1)) so that the ' equivalent ' temperature T_0 is a convenient way of expressing the minimum of the potential energy curve of fig. 4.2.

	Ne	Ar	Kr	Xe	N_2
σ in Å	2·75	3·405	3·60	4·10	3·70
$T_0 = \epsilon/k$ in K	35·60	119·800	171·00	221·00	95·10

Table 4.1. Constants in the 12–6 function (equation (4.10)).

(This table is reproduced from *The Liquid State* by J. A. Pryde.)

4.6. *Other simple forms of the potential energy function*

We now mention three other, simpler forms of the function ϕ which have been used in the study of liquids.

In the first of these each molecule is regarded as a perfectly rigid non-attracting sphere of diameter σ. This so-called ' rigid-sphere ' potential energy is then defined by

$$\phi = \infty \quad \text{for } r < \sigma$$

and

$$\phi = 0 \quad \text{for } r > \sigma.$$

A more realistic model is to consider rigid spheres which exert attractive forces on each other. In this case we have

$$\phi = \infty \text{ for } r < \sigma$$

and

$$\phi = -A/r^6 \text{ for } r > \sigma.$$

(This attractive energy is of the type given in equation (4.4).)

A third type of function is the 'square-well' potential energy function given by

$$\begin{aligned} \phi &= \infty \quad &\text{for } r < \sigma \\ \phi &= -\epsilon_1 \quad &\text{for } \sigma < r < \sigma_1 \\ \phi &= 0 \quad &\text{for } r > \sigma_1 \end{aligned}$$

where σ_1 is a length greater than the diameter σ and ϵ_1 is an energy; σ_1 is usually taken to be about $1{\cdot}5\sigma$.

These types of potential energy function have been used a good deal in theoretical work and are shown in fig. 4.6. For example in the

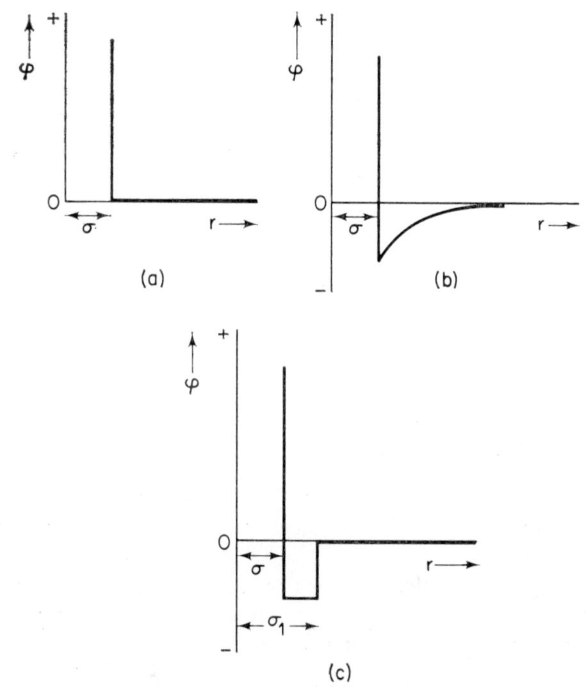

Fig. 4.6. Graphs showing (*a*) the rigid-sphere potential energy function, (*b*) the energy function for rigid spheres with attractive forces, and (*c*) the square-well function.

Monte Carlo and molecular dynamics methods of studying assemblies of molecules the rigid sphere and square-well functions have been used. These methods will be discussed in Chapter 5.

4.7. *The use of statistical mechanics*

We now discuss how the methods of statistical mechanics can be used to treat an assembly of N molecules comprising a liquid. To do this we first consider a perfect gas of N molecules, each of mass m, in an enclosure of volume V at a temperature T. If the permissible energies for any molecule are $\epsilon_1, \epsilon_2 \ldots, \epsilon_i, \ldots$ then the *partition function* for any molecule is defined, in general, as

$$f = \sum_i \exp\left(-\epsilon_i/kT\right). \tag{4.13}$$

For one molecule of the perfect gas under consideration it can be readily shown that

$$f = V\left(\frac{2\pi m\, kT}{h^2}\right)^{3/2} \tag{4.14}$$

where k is the Boltzmann constant and h is the Planck constant. For the whole perfect gas assembly the partition function Z is given by

$$Z = \frac{f^N}{N!} = \frac{V^N}{N!}\left(\frac{2\pi m\, kT}{h^2}\right)^{3N/2}. \tag{4.15}$$

In a perfect gas, as we saw in section 3.1, there are no interactions between the molecules and the above equations have taken account of the *translational* kinetic energies of the molecules only. But in an assembly where there *are* interactions between the molecules—such as in an imperfect gas or liquid—these interactions must be accounted for by including an extra term in the partition function (4.15). So we rewrite (4.15) as

$$Z = \left(\frac{2\pi m\, kT}{h^2}\right)^{3N/2} \underset{\sim}{Q} \tag{4.16}$$

where $\underset{\sim}{Q}$ includes the term $V^N/N!$ of (4.15) and also the effects of the interactions. The contribution from these interactions occurs in the form of a term $\exp(-\Phi/kT)$ where Φ is the total potential energy of the assumbly of N molecules.

We calculate Φ as follows. Suppose that the N molecules are labelled $1, 2, \ldots i, j, \ldots N$ and that the mutual potential energy of molecules i and j is ϕ_{ij}. We then assume that the molecules interact in pairs only. Then the total potential energy of the assumbly is

$$\Phi = \sum \phi_{ij} \tag{4.17}$$

where the summation extends over all the distinct ij pairs of molecules.

The number of such pairs is $NC_2 = N(N-1)/2$. We see therefore the importance in theoretical work of the mutual potential energy function ϕ. In principle, if the ϕ_{ij}'s in (4.17) are known then Φ and hence Q can be determined. Indeed the crux of the problem is to evaluate this ' configurational ' partition function Q and this has been the great achievement of computer calculations in recent years. Once Q is known the partition function Z is known.

This partition function Z is one of the most important functions in statistical mechanics. If we know Z the other thermodynamic properties of the assembly follow at once. For example, the internal energy U of the assembly is

$$U = \frac{kT^2}{Z}\left(\frac{\partial Z}{\partial T}\right)_V$$

and the equation of state is

$$p = kT\left(\frac{\partial \ln Z}{\partial V}\right)_T.$$

References

(1) Pryde, J. A., *The Liquid State*, Chapter 4 (Hutchinson University Library, London, 1966).

(2) Flowers, B. H., and Mendoza, E., *Properties of Matter*, Chapter 3 (John Wiley & Sons Ltd., 1970).

(3) Tabor, D., *Gases, Liquids and Solids*, Chapter 2 (Penguin Library of Physical Sciences, 1969).

CHAPTER 5

various methods of studying the structure of a liquid

5.1. *Introduction*

In this chapter we summarize some of the methods which have been used to study the structure of a liquid on the molecular scale. We shall first consider the more theoretical approaches; these are the Monte Carlo method, the molecular dynamics method and the method based on the use of integral equations. Secondly we consider the results which have been obtained from studying assemblies of plasticine spheres and steel ball-bearings. Finally we describe diffraction experiments which have been carried out using the actual liquids themselves.

5.2. *The Monte Carlo method*

At any instant of time the molecules in a liquid will have a certain spatial configuration. As time goes on this spatial configuration changes continuously. This is the basis of the Monte Carlo method. We start off with an assembly of a small number of molecules in some initial spatial configuration and then we generate a subsequent sequence of configurations by moving the molecules one at a time, thus simulating the effect of thermal agitation. Then we average the particular property in which we are interested over all the chosen configurations, a method known as 'ensemble averaging'.

In practice this sequence of configurations is calculated by means of a large and fast computer, starting with a configuration in which the molecules are in some regular lattice arrangement. Even with such modern computers we can only cope with 'assemblies' of the order of a hundred molecules. Calculations for such small assemblies of rigid spheres and of molecules interacting with 12–6 potentials have been made. At different stages in the history of the assembly we can calculate such things as the radial distribution function, the mean kinetic energy per molecule or the pressure. The properties of these very small assemblies are not so very different from those of larger ones; hence if we work with assemblies of different sizes it is possible to extrapolate the results to real assemblies having very large numbers of molecules.

5.3. *The molecular dynamics method*

This method again involves the use of computers and is in some ways more ambitious than the Monte Carlo method. We begin with a number of molecules, N, in a lattice arrangement in a box; the molecules are then started off with equal speeds but with random directions of motion. The subsequent changes in the assembly with time are followed by solving the individual equations of motion of all the molecules. Even for a small numbei of molecules this involves a very large computer programme. It is found that, after only a relatively small number of collisions between the molecules, the velocity distribution is virtually the Maxwell distribution. If the molecules are rigid spheres their energies are entirely their kinetic energies and the total energy U of the assembly depends on the initial speeds given to the molecules; U does not change with time and it also determines the effective temperature of the assembly since $U=(3/2)NkT$ (see equation (3.1)). ' Periodic ' boundary conditions are chosen so that a molecule passing out of one side of the box re-enters with the same velocity through the opposite side; this ensures that the number of molecules remains constant at N. The equilibrium value of any particular property is found by time averaging and quantities like the pressure and radial distribution function can be found. Alder and Wainwright used these methods to study an assembly of 32 rigid spheres and later one of 108 molecules with square-well potentials (section 4.6). The computer programme enabled the dynamical history of each molecule to be recorded.

The motion of the molecules depends on their initial speeds and on their tightness of packing. If the motion consists of vibrations about well-defined ' sites ' then this corresponds to a crystalline solid. If the motion is more energetic the molecules vibrate with larger amplitudes about these sites and sometimes ' swop ' sites and diffuse slowly from place to place; this corresponds to the liquid phase. There is a third possibility in which the molecules move around throughout the whole box and this represents the gaseous or vapour phase.

In fig. 5.1 (a) and (b) some results obtained by Alder and Wainwright for an assembly of 32 rigid spheres are shown. The tracks in these photographs are the projections of the molecular trajectories on to one face of the box. In fig. 5.1 (a) the trajectories show small motions about well-defined sites and this clearly represents the ' solid ' phase. Fig. 5.1 (b) shows a ' liquid ' phase in which most of the molecules are moving within a ' cell ' of neighbouring molecules but they also swop places occasionally. In fig. 5.1 (c) the projected trajectories for a less closely packed assembly of 108 molecules with square-well potentials are shown; this molecular behaviour is similar to the ' vapour ' phase Fig. 5.1 (a) and (b) show clearly the differences and similarities between the molecular motions in the solid and liquid phases.

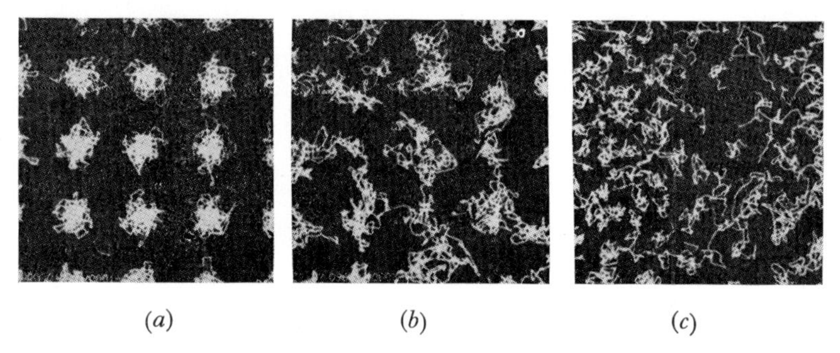

(a) (b) (c)

Fig. 5.1. (a) Projected trajectories for 32 rigid spheres illustrating the 'solid' phase (after Alder and Wainwright). (b) Projected trajectories for 32 rigid spheres illustrating the 'liquid' phase (after Alder and Wainwright). (c) Projected trajectories for 108 molecules with square-well potentials illustrating the 'vapour' phase (after Alder and Wainwright).

5.4. *The use of integral equations*

We have already discussed the radial distribution function $g(r)$ in section 2.3. There we chose the centre of a certain molecule as the origin of coordinates and we saw that the other molecules were situated in 'shells' surrounding this chosen molecule. Thus as we travel outwards from the origin the local number density is not constant but oscillates about its mean value as shown in the curve of fig. 2.2. As we shall see in section 5.7 X-ray and neutron diffraction work confirm that the molecules are arranged spatially in the way shown by this curve, with certain distances from the origin being more likely than others due to the geometrical constraints imposed by filling up the space with spherical molecules.

Kirkwood saw that it was possible to calculate $g(r)$ mathematically from first principles and he did this by means of an *integral equation.* We imagine building up a liquid, shell by shell, around a given central molecule. In placing a molecule in a given position we first ask whether it is too near our central molecule; if it is not we ask further whether its proposed position is already occupied or is impossible because of the proximity of other molecules. So we have to examine each point of our space in this way and this explains why *integrations* come into the theory. The mathematical methods involved are very difficult and the most successful form of integral equation is that given by Percus and Yevick in 1958. It enables the function $g(r)$ to be predicted theoretically to a degree of accuracy comparable with that given by experimental methods. The importance of $g(r)$ lies in the fact that the thermodynamic properties of the liquid can be expressed in terms of $g(r)$ and the intermolecular potential energy ϕ.

One other point is worth mentioning. The radial distribution function for any liquid, no matter what the intermolecular interaction is like, is very similar to that which would be expected if the molecules were rigid spheres. In other words the form of $g(r)$ is determined largely by *geometrical* considerations and the fact that *real* molecules have attractions and are not absolutely ' rigid ' spheres does not affect it appreciably. This leads us naturally to the geometrical models of the next section.

5.5. *Three-dimensional models*

Some very interesting and important results have been obtained by Bernal and his colleagues on the random packing of spheres.

In one set of experiments a number of plasticine spheres were placed in a rubber bladder after being previously chalked to prevent them from sticking. After removing any enclosed air to ensure that there were no air bubbles the spheres were compressed together so that they filled the whole of the space inside the bladder. The contents of the bladder were then examined and this showed that the compressed spheres had become polyhedra, the average number of faces being 13·6. Most of the polyhedral faces were five-sided. Bernal then used geometrical arguments to show that in a heap of such polyhedra with predominantly five-fold faces there will be a preponderance of five-fold ' rings ' of polyhedra; this must lead to irregular packing since this five-fold symmetry cannot correspond to a repeating structure in space and so cannot lead to a *long-range* three-dimensional arrangement. (Results from crystallography show that we can have 2-fold, 3-fold, 4-fold or 6-fold symmetry but certainly not 5-fold.) Bernal states that " the inevitable appearance of 5-fold rings eliminates any possibility of long-range order ".

The random packing of steel ball bearings was also studied by placing about three thousand $\frac{1}{4}$ inch bearings into a rubber balloon which was then squeezed and shaken so that random close packing of maximum density was obtained. Paint was then poured in and the whole assembly allowed to dry; the balloon was then removed. The structure was then broken up and the number of contacts and near-contacts for some 500 balls was studied. A part of the dismantled assembly of spheres is shown in figs. 5.2 (I) and 5.2 (II); the dots on the balls indicate where balls had nearly touched and the rings show where they had exactly touched.

The results showed that the *average* coordination number in the ' liquid ' assembly was about 9 compared with 12 for a closed-packed solid. If we assume that the nearest-neighbour separation in liquid and solid are about equal this shows how the greater specific volume of a liquid (molten solid) can be explained by a decrease in the coordination number (see section 2.2). It was also possible to determine the radial distribution function of the assembly of ball-bearings; this follows quite closely the *experimental* distribution function for liquid argon obtained from neutron diffraction work. This shows that the radial distribution function of a simple monatomic liquid like argon is essentially that for random close packing of an assembly of steel balls.

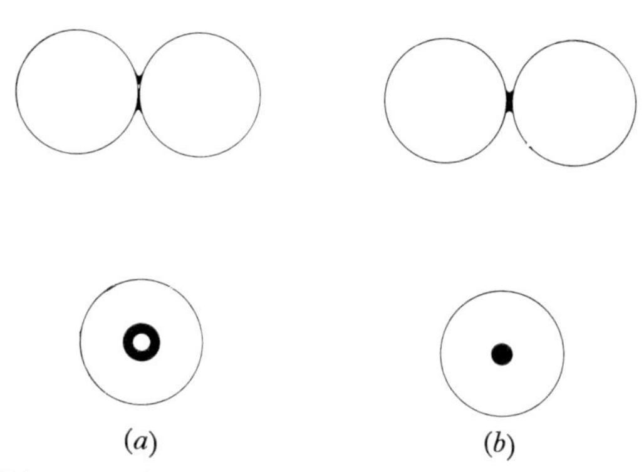

Fig. 5.2 (I). Diagram of method of marking (*a*) close and (*b*) near contacts between spheres. The areas of adherent black paint are marked.

Fig. 5.2 (II). Portion of random close packed ball assembly showing marks of further contacts.

(Fig. 5.2 is reproduced with the permission of the Editor of *Nature*.)

Scott and his colleagues have carried out similar experiments with steel ball bearings using molten paraffin wax to ' bind ' the assembly of balls. The radial distribution function was again obtained. In Fig. 5.3 the results obtained by Bernal and Scott are compared with

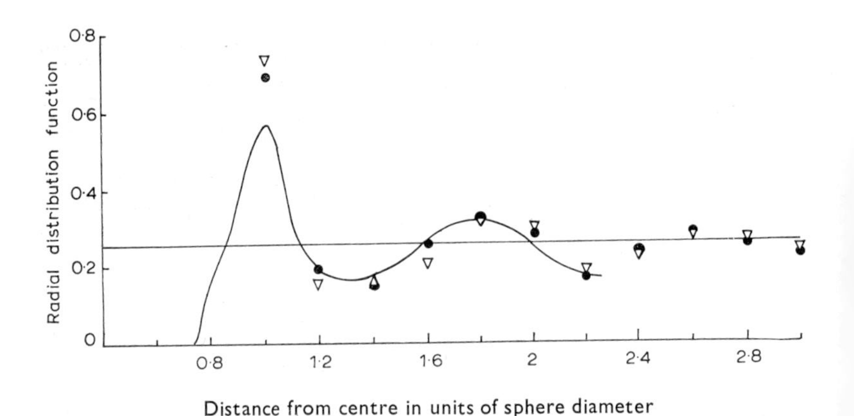

Fig. 5.3. Radial distribution function for liquid argon (shown as the continuous curve) together with the results of Bernal (●) and Scott (▽).

the experimental radial distribution function for liquid argon derived from neutron diffraction experiments.

The work of Bernal and of Scott is essentially concerned with the *geometrical* structure of a liquid. Indeed Bernal refers to it as 'statistical geometry'.

5.6. *Two-dimensional models*

In the previous section we considered a three-dimensional aggregate of steel balls bound together by either paint or paraffin wax. Such a static model really represents a 'flash photograph' of a liquid at a given instant and it does not allow for the effect of the molecular thermal motions to be examined. Neither does it enable the effect of intermolecular *attractions* to be studied since the steel balls behave as 'rigid spheres'. An attempt to study these two effects using a two-dimensional model has been recently made by Walton and Woodruff. In this model a number of ball-bearings were placed on a rough-moulded glass tray. The 'attractions' were provided by coating the ball-bearings with oil whilst the 'thermal motion' was produced by vibrating the tray by means of an eccentric drive from a motor. This model thus enabled the behaviour of a two-dimensional 'monatomic liquid' to be examined.

Initially the oil-covered bearings were placed on the *stationary* tray such that they formed a regular close-packed two-dimensional lattice with a coordination number of $z=6$; this represents the 'solid'. The tray was then set in vibration and as the energy of this vibration (the 'temperature') was increased the lattice remained unchanged

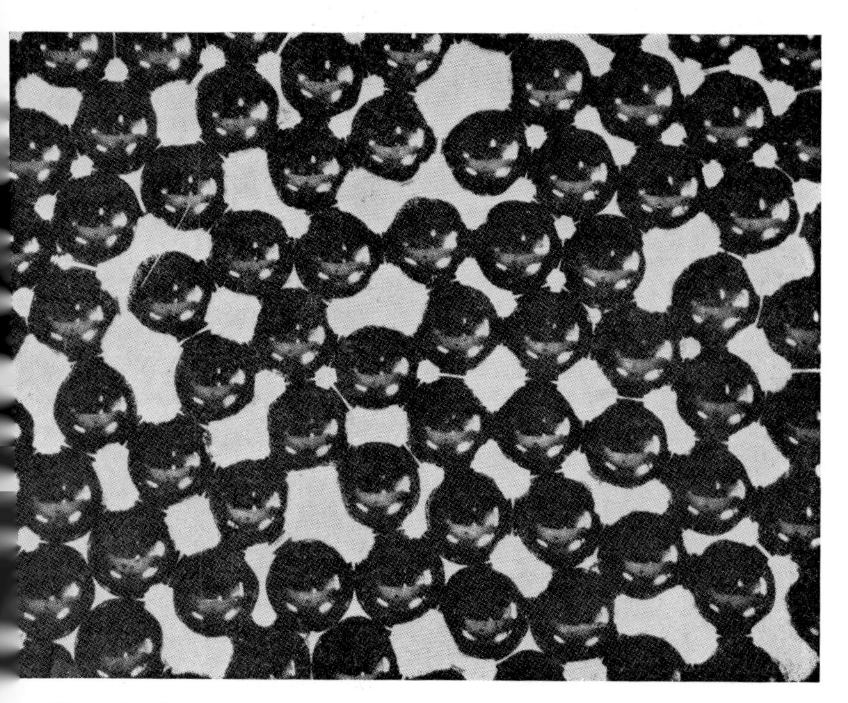

Fig. 5.4. An instantaneous ' snapshot ' of a simulated two-dimensional ' liquid ' (after Walton and Woodruff).

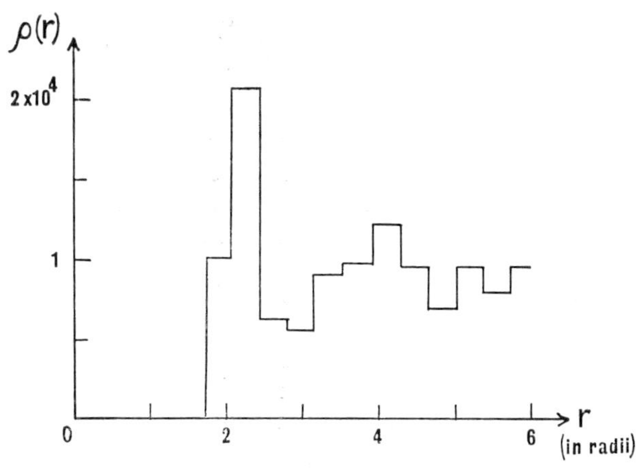

Fig. 5.5. Histogram showing the radial variation in density $\rho(r)$ (of centres) about any one bearing (after Walton and Woodruff).

until at a certain 'temperature' the 'solid' assembly melted into a 'liquid'. A still from a cine film of the changes occurring is shown in fig. 5.4 and this shows the 'liquid' phase after the 'melting'. This frame shows how each ball in the 'liquid' is surrounded, on average, by five nearest neighbours and not six as in the 'solid'. This is a further illustration of the fact that the number of nearest neighbours in a liquid is less than that in a solid. The sequence of cine pictures also shows that any one ball is confined within its cell of nearest neighbours for most of its time.

In further measurements on these photographs Walton and Woodruff obtained the radial distribution function for the two-dimensional 'liquid' and this showed the general pattern of peaks and troughs that one would expect (fig. 5.5; cf. fig. 2.2).

5.7. *Results from X-ray and neutron diffraction*

We now turn to some diffraction experiments which have been performed on the actual liquids themselves.

Diffraction effects of this kind, as usually observed, result from the scattering of a collimated beam of waves (of wavelength λ) by 'obstacles' of dimensions comparable with λ arranged in a pattern with a characteristic repeating interval d. The scattered waves reinforce to give strong 'reflections' or 'transmissions' (which we can call diffracted beams) at certain angles θ to the incident beam and the possible values of θ depend on d and λ. If an identified θ is measured and λ is known, d can be determined.

For example, with the optical diffraction grating, d is the spacing o the rulings and $n\lambda = d \sin \theta$ where n is zero or a (usually) smal integer. In the Bragg crystal X-ray spectrometer d is the separatio between adjacent lattice planes and $n\lambda = 2d \sin \theta$. In the early experi ments a beam of monochromatic X-rays (for which λ is of the orde of 10^{-10} m) was allowed to fall on a single crystal and the diffractio picture produced on a photographic plate on the other side of th crystal was found to be a set of regular discrete spots surrounding a intense central spot produced by the undeviated ray. The sharpnes and discreteness of the pattern is due to the long-range order existin in the single crystal.

Next, suppose that the long-range order is removed by powderin the crystal thus transforming it into a large number of tiny crysta orientated in all possible directions. The X-ray picture now obtaine does not show discrete spots but a series of well defined rin surrounding the central spot. It is a short step from a powdere crystal to a liquid, and in 1916 Debye and Scherrer studied t picture obtained when X-rays were passed through a liquid. Th picture again showed the central spot and a set of rings which we less well defined. The existence of these rings shows that there is short range order in the liquid but the absence of any discrete spots indicates that there is no long-range order present.

Briefly, the radial distribution function for a liquid may be derived from X-ray diffraction measurements in the following way. The intensity I of the diffracted light is obtained as a function of the angle θ between the undeviated and diffracted beam. This intensity can be measured photographically but more recently this has been done using Geiger counter methods. The value of the radial distribution function $g(r)$ can then be obtained from the measured value of I using a mathematical technique known as Fourier inversion. The use of modern computers makes this technique a fairly easy matter. Examples of radial distribution functions obtained from X-ray diffraction were given in fig. 2.3.

We mentioned earlier that radiation whose wavelength was about 10^{-10} m was suitable for these experiments. A beam of neutrons can ılso behave as radiation having a wavelength of this order. This is because a material particle of mass m moving with a velocity v has ı wavelength of $\lambda = h/mv$, where h is the Planck constant, associated with it. Experiments using a monochromatic beam of neutrons have been carried out by a number of workers and we have already given ın example in fig. 5.3 of a typical distribution function obtained rom neutron diffraction. The methods of neutron diffraction have ıot been as widely used as those involving X-rays. Neutron diffraction has however certain advantages denied to X-rays and can yield nore overall information. For example neutrons are scattered by he *nuclei* of atoms (as distinct from X-rays which are scattered by the lectrons). Thus neutrons are scattered by the light elements such s hydrogen, helium, etc. For such light elements X-ray scattering virtually negligible. An excellent account of neutron scattering by quids has been given by J. E. Enderby (see p. 106).

.8. *Conclusion*

On the whole there is pretty good agreement between the various ıeoretical and experimental methods of measuring the radial istribution function for a liquid. Possibly the most interesting oint is that the experimental distribution function for liquid argon, otained from neutron diffraction, agrees so well with that obtained y Bernal and by Scott using their ‘ rigid ’ ball-bearings.

Finally it should be mentioned that laser light diffraction and trasonic methods can be used to study liquids near the critical point here density fluctuations occur over distances of the order of a ousand intermolecular separations. Such wavelengths are then ıitable.

References

(1) PRYDE, J. A., *The Liquid State*, Chapters 3 and 8 (Hutchinson University Library, London, 1966).
(2) TABOR, D., *Gases, Liquids and Solids*, Chapter 11 (Penguin Library of Physical Sciences, 1969).
(3) BARKER, J. A., *Lattice Theories of the Liquid State*, Chapter 2 (Pergamon Press, 1963).
(4) WALTON, A. J., and WOODRUFF, A. G., Kinetic Liquid-Simulator, *Contemp. Phys.*, **10,** 59 (1969).
(5) 'Neutron Scattering in Liquids', by ENDERBY, J. E. (see reference on p. 106).
(6) BACON, G. E., and NOAKES, G. R., *Neutron Physics* (Wykeham Publications (London) Ltd., 1969).

CHAPTER 6

phase changes

6.1. *Some results from thermodynamics*

We have already met certain thermodynamic properties such as pressure (p), absolute thermodynamic temperature (T), volume (V) and internal energy (U). We now introduce a few more such functions together with some important relations which they satisfy.

Consider a system in which we have a fixed quantity of matter in equilibrium at an absolute temperature T and pressure p. (Such a system could be a gas of N molecules in its enclosure of volume V). Now suppose that this system absorbs a small quantity of heat $\mathrm{d}Q$ at the temperature T. The system is then transformed from one equilibrium state to another. This heat $\mathrm{d}Q$ is used up in two ways, namely, to increase the internal energy of the system by $\mathrm{d}U$ and to do external work of amount $p\mathrm{d}V$ as the volume expands by $\mathrm{d}V$. So we write

$$\mathrm{d}Q=\mathrm{d}U+p\mathrm{d}V \tag{6.1}$$

which is the first law of thermodynamics.

We assume further that the process of heat absorption by the system is reversible. By this we mean that the system, in its final equilibrium state, can *give up* a quantity of heat $\mathrm{d}Q$ at the temperature T and so revert to its initial state. The reversible absorption of heat $\mathrm{d}Q$ at this temperature T causes a particular property of the system, known as its entropy S, to increase also. This increase of entropy $\mathrm{d}S$ is related to $\mathrm{d}Q$ by the relation

$$\mathrm{d}Q=T\mathrm{d}S \tag{6.2}$$

which follows from the second law of thermodynamics. In equation (6.2) we are defining a *change* $\mathrm{d}S$ in the entropy rather than its absolute value.

From equations (6.1) and (6.2) we have

$$T\mathrm{d}S=\mathrm{d}U+p\mathrm{d}V. \tag{6.3}$$

We now define two further thermodynamic quantities. These are the Helmholtz free energy F for changes at constant volume and temperature and the Gibbs free energy G for changes at constant pressure and temperature. They are defined thus:

$$F = U - TS \tag{6.4}$$

$$G = U + pV - TS. \tag{6.5}$$

Let us now concentrate on the Gibbs function G. A small change $\mathrm{d}G$ in G is given by

$$\mathrm{d}G = \mathrm{d}U + p\mathrm{d}V + V\mathrm{d}p - T\mathrm{d}S - S\mathrm{d}T$$

which, if we also use equation (6.3), reduces to

$$\mathrm{d}G = V\mathrm{d}p - S\mathrm{d}T. \tag{6.6}$$

In a reversible change which occurs at constant temperature and pressure, $\mathrm{d}T = 0$ and $\mathrm{d}p = 0$ and so $\mathrm{d}G = 0$; thus the Gibbs function for a given quantity of matter (usually we take unit mass or one mole) remains a constant in such a change. The transitions of melting and vaporization, which we briefly considered in sections 1.2 and 1.3 are examples of such reversible changes. So if two phases of the same substance coexist in equilibrium their Gibbs functions per unit mass must be equal.

6.2. *The Clapeyron equation*

Referring back to fig. 1.2 let us consider the liquid–vapour equilibrium line RC only as shown in fig. 6.1 below.

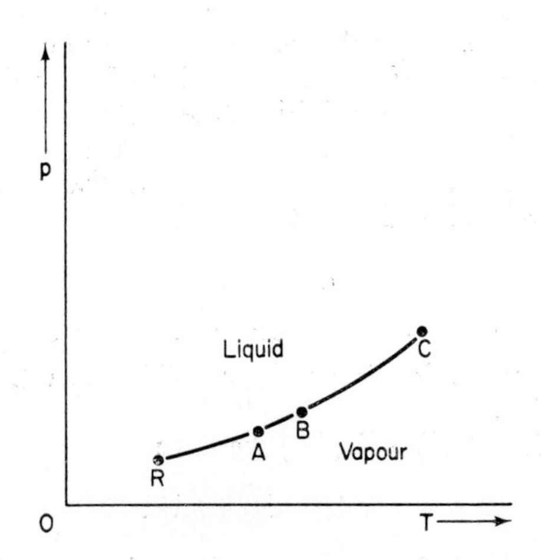

Fig. 6.1. RC is the liquid–vapour equilibrium line and A,B are two neighbouring points on this line.

Consider unit mass of a substance and let the suffices 1 and 2 refer to the liquid and vapour phases respectively. At the point A, corresponding to a pressure p and temperature T, the Gibbs function G is the same for unit mass of the liquid and unit mass of the vapour in equilibrium with it. So we write

$$G_1 = G_2.$$

At a neighbouring point B on the equilibrium curve where the pressure is $p + \mathrm{d}p$ and the temperature $T + \mathrm{d}T$ we similarly have

$$G_1 + \mathrm{d}G_1 = G_2 + \mathrm{d}G_2.$$

Subtracting the last two equations gives us

$$\mathrm{d}G_1 = \mathrm{d}G_2$$

which, using equation (6.6) with appropriate suffices, leads to

$$V_1 \mathrm{d}p - S_1 \mathrm{d}T = V_2 \mathrm{d}p - S_2 \mathrm{d}T.$$

Thus

$$\frac{\mathrm{d}p}{\mathrm{d}T} = \frac{l_{\mathrm{v}}}{T(V_2 - V_1)} \tag{6.7}$$

where $l_{\mathrm{v}} = T(S_2 - S_1) = T\Delta S$ is the specific latent heat of vaporization. This amount of heat is that which must be absorbed by unit mass of the liquid at the temperature T in order to change it into the vapour phase. Equation (6.7) is known as the Clapeyron equation. Since V_2 is always greater than V_1 then $\mathrm{d}p/\mathrm{d}T$ is always positive.

If we similarly consider the melting curve RA of fig. 1.2 we can derive the result

$$\frac{\mathrm{d}p}{\mathrm{d}T} = \frac{l_{\mathrm{f}}}{T(V_2 - V_1)} \tag{6.8}$$

by the same method, where l_{f} is the specific latent heat of fusion and V_1, V_2 are now the volumes of unit mass of the solid and liquid phases respectively. Since most substances expand on melting then $V_2 > V_1$ and $\mathrm{d}p/\mathrm{d}T$ is positive so that increasing the pressure raises the melting point. Ice and bismuth are exceptions to this general rule and for these substances $V_2 < V_1$, $\mathrm{d}p/\mathrm{d}T$ is negative and an increase in pressure causes a lowering of the melting point.

The two phase transitions of vaporization and melting are each a so-called *first order* transition. During each transition the pressure, temperature and Gibbs function per unit mass remain unchanged but the values of the specific volume and specific entropy jump by finite amounts of $\Delta V = V_2 - V_1$ and $\Delta S = S_2 - S_1$ respectively.

6.3. *Theories of evaporation and condensation*

We now discuss briefly the theories which have been proposed to explain evaporation and the converse process of condensation.

First we have the explanation given by the van der Waals equation. This was discussed in some detail in sections 3.2 to 3.5.

The second type of theory is that based on the Mayers' elegant theory of imperfect gases. To illustrate this we start by considering a gas of fairly low density consisting of N molecules in a volume V; in this case there are negligible interactions between most of the molecules since they are fairly far apart and the state of affairs approximates to that of a perfect gas. Suppose now that this gas is compressed and its density thereby increased. As the molecules get closer together the interactions between them are no longer negligible and the gas becomes an 'imperfect' gas. The Mayers pictured this imperfect gas as being made up of clusters containing 2, 3, . . . , l, . . . molecules; the molecules in any particular cluster interact with one another because of their mutual proximity. In this case the equation of state turns out to be

$$\frac{p}{kT} = \sum_l b_l Z^l \tag{6.9}$$

where b_2, b_3, . . . , b_l, . . . are cluster integrals corresponding to clusters of 2, 3, . . . , l, . . . molecules and Z is a quantity known as the fugacity which satisfies the relation

$$\frac{N}{V} = \sum_l l b_l Z^l. \tag{6.10}$$

The value of b_1 is defined to be unity.

For the perfect gas case the only term in either (6.9) or (6.10) is the term for $l=1$ as there are no clusters of $l>1$ in a perfect gas. So in this case (6.9) and (6.10) reduce to

$$\frac{p}{kT} = Z$$

and

$$\frac{N}{V} = Z.$$

These together give $pV = NkT$ which is the perfect gas law which holds at very low densities. For rather higher densities the clusters mainly responsible for the departure of the gas from perfect behaviour are the small ones which are initially formed; these clusters contain 2, 3, 4, etc. molecules. But as the density continues to increase a stage is reached where a very sudden formation of *large* clusters takes place. This is the condensation point. In this way it is possible to trace the stages by which an initially perfect gas proceeds to ecome an imperfect gas as its density increases and finally, as this ensity continues to increase, to condense.

.4. *Theories of melting*

In spite of many attempts there is still no adequate theory of nelting. Perhaps the best known is that due to Lindemann who iscussed the changes occurring as a crystalline solid with an ordered nolecular arrangement melts into a less ordered liquid. The solid ttice is kept together by the attractive forces between each molecule nd its neighbours. The individual molecules vibrate about their ttice sites with amplitudes which will continually increase if the emperature is steadily raised. Lindemann assumed that melting ccurs when the amplitude of vibration exceeds a certain critical raction of the distance between nearest-neighbour molecules in the olid lattice. At melting the long-range order is destroyed and only ome local order is preserved. For further details see Tabor's book *Gases, Liquids and Solids*, Chapter 11.

.5. *Liquid crystals*

In sections (6.2) and (6.4) we discussed the usual melting of a rystalline solid into a liquid. There are however certain crystalline olids which when heated do not change at once into an isotropic quid but pass first through an intermediate phase known as a *nesophase.* This phenomenon was first observed by Reinitzer in 888. He found that if solid cholesteryl benzoate was heated it hanged first into a turbid liquid at 145°C and later into a clear liquid t 179°C. Similar behaviour in other substances has been observed y Lehmann who has demonstrated that the intermediate turbid quid or mesophase is birefringent (double-refracting). Birefringence s the splitting of a ray of unpolarized incident light into two refracted ays plane polarized in planes at right angles to each other and is a roperty one associates with a crystal. So the mesophase resembles liquid in some ways and a crystal in other ways; for these reasons it is ften called a *liquid crystal.* For these substances two transition emperatures are involved. First there is a temperature T_1 at which ne crystalline solid melts into the mesophase and a second higher emperature T_2 at which the muddy mesophase changes into a clear otropic liquid (fig. 6.2).

Substances which form liquid crystals usually have long rod-like nolecules. In the solid crystalline phase these molecules form an rrangement in which there is high spatial and angular correlation. n the mesophase the molecules tend to set themselves with their ng axes parallel so that the angular correlation, but not the marked patial correlation, of the solid still exists. Finally, in the isotropic quid phase, the angular correlation also disappears.

Let us now consider the mesophase in greater detail. In additio
to the high angular correlation of the molecules there are other variou
possible structural factors which lead to different types of mesophase
Three basic types have been identified and are known as *nematic*
cholesteric and *smectic* mesophases. A nematic mesophase i
structurally the simplest and it has a low viscosity. A cholesteri
mesophase is produced by many esters of cholesterol; its viscosit
is higher than that of the nematic mesophase and it is optically active
A smectic mesophase is much more like a solid in its structure an
has a far higher degree of order than the other two.

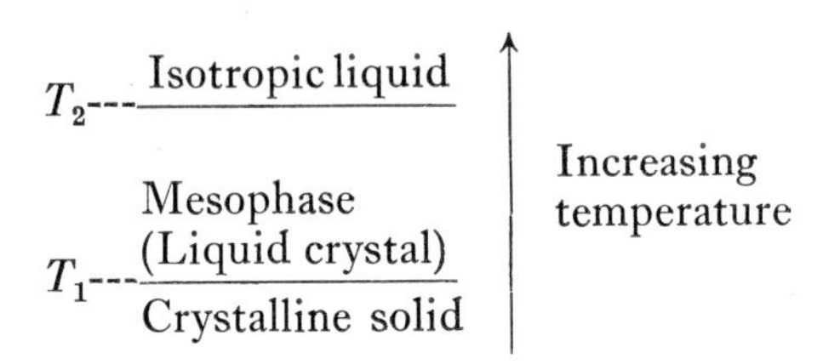

Fig. 6.2. Illustrating the two transition temperatures associated with a liquid crystal.

In a nematic liquid crystal the axes of the long molecules tend t
arrange themselves parallel to one another but there is no correlatio
between the molecular centres. This is really an over-simplifie
description because the long axes are by no means completely alignec
The detailed description of the arrangement involves the notion of
'preferred' molecular orientation at a given point in the crystal an
this particular orientation is described by a vector known as a *directc*
(de Gennes, 1969). The changes in the orientation of the directc
from place to place inside a nematic mesophase are the cause of th
turbid appearance of the phase. 'Imperfections' in a nematic liqui
crystal can arise just as in most solid crystals and manifest themselve
as *disclination* lines which appear as threads when observed in
microscope.

The cholesteric mesophase is really a special case of a nemati
liquid crystal. The molecules in a cholesteric mesophase again hav
a tendency to set themselves with their long axes parallel. If
cholesteric mesophase is sandwiched between two glass surfaces th
molecules are arranged in sheets; two such adjacent sheets A and
are shown in fig. 6.3. As we go from sheet A to sheet B the directc
is turned through a certain angle about an axis Z′Z perpendicular t
the sheets. The next sheet C (not shown) above B has a directc
turned through the same angle with respect to that of B and so o
until eventually we reach a sheet in which the director is in the sam
direction as that in sheet A. Thus the structure has a helica

repeating pattern and this helical nature of the director has been detected using optical polarization methods.

In smectic liquid crystals the centres of the molecules are highly correlated in addition to the molecular long axes being parallel and this mesophase has much in common with a solid.

The number and variety of technological applications for liquid crystals are very large and will undoubtedly grow in the future. To mention only a few examples, liquid crystals have been used in (*a*) electronic display systems, wrist watch faces, portable calculating machines and (*b*) medical thermography, thermal nondestructive tests of aerospace components and the detection of structural flaws in integrated circuits.

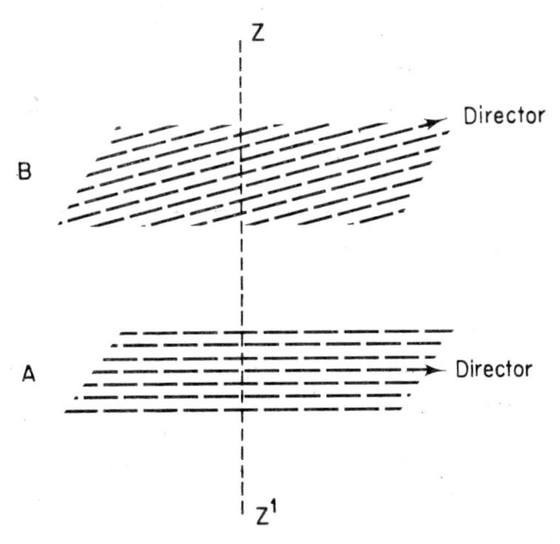

Fig. 6.3. The helical structure in a cholesteric mesophase showing the difference in the orientation of the directors in sheets A and B.

Nematic liquid crystals are used in the type (*a*) applications just mentioned. The display system consists of a nematic liquid crystal sandwiched between two plates of glass each having a transparent conducting coating. If there is no applied voltage across the plates the enclosed mesophase appears transparent. The application of a small d.c. or low-frequency a.c. voltage however results in the creation of a number of scattering centres in the mesophase; in other words the mesophase scatters incident light, a process known as an *electro-optic* effect. If one of the plates is made up of a number of segments, each photoetched with a number (0 to 9) or a letter, then the required number or letter may be produced by applying the voltage to the particular segment concerned. In this way numerical displays for electronic clocks, watches and other digital instruments can be produced.

Cholesteric liquid crystals are used in the type (*b*) applications mentioned previously. In a cholesteric mesophase the 'pitch' of the helical pattern can be caused to change by even very small changes in the temperature. If the pitch changes this will produce changes in the wavelength of any light reflected by the mesophase. Thus the abrupt changes in colour of this type of mesophase due to small temperature changes gives us a good visual temperature-sensing detector. These methods have been employed to detect structural faults in electronic and aerospace components and in turbine blades. In medicine they have been used to monitor skin temperatures over a fairly large area of the body surface and thus detect such things as breast tumours, where the temperature would be higher than that of the surrounding area. The method involves applying a light-absorbing coating to the area to be investigated and covering this coating with a calibrated cholesteric liquid crystal. This liquid crystal is designed to respond to a temperature variation of about 4 kelvin over the area examined; the coldest regions appear red or orange and the warmer ones blue or violet. By directly illuminating the area the patient can be examined visually and a colour photograph can be taken as a permanent record.

6.6. *Glasses*

A certain class of liquids can be cooled below their normal melting points and such a supercooled liquid hardens into a 'glass' after a time. Although we tend to think of a piece of ordinary glass as being a solid the molecular arrangement in a glass closely resembles that in a liquid. There is short-range order but no long-range order. Indeed X-ray diffraction experiments on glasses produce the pattern of diffuse rings which we associate with a liquid (see section 5.7). One can regard the molecular structure in a glass as being an instantaneous photograph of that in a liquid. The glassy state is also like a liquid in that it flows very slowly indeed when stressed indicating that it behaves like a liquid of abnormally high viscosity.

References

(1) Pippard, A. B., *The Elements of Classical Thermodynamics* (Cambridge University Press, 1964).

(2) Zemansky, M. W., *Heat and Thermodynamics*, 5th edition (McGraw-Hill Book Co., 1968).

(3) Temperley, H. N. V., *Changes of State* (Cleaver-Hume Press Ltd., London, 1956).

(4) Tabor, D., *Gases, Liquids and Solids* (Penguin Library of Physical Sciences, 1969).

(5) 'Liquid Crystals'. Article by G. R. Luckhurst in *Physics Bulletin* **23,** 279 (1972).

(6) De Gennes, P. G., 1969, *Mol. Cryst. Liq. Cryst.*, **7,** 325.

(7) *Liquid Crystal Devices*, edited by Thomas Kallard (Optosonic Press, New York, 1973).

CHAPTER 7

the surface layer

7.1. *Introduction*

We have already discussed how short-range order exists around any molecule in the interior of a liquid (section 2.2). But a molecule which is fairly near the surface of the liquid will have a different and more unsymmetrical environment because below it is the bulk liquid and above it a more rarefied assembly of vapour molecules. For these reasons we expect the surface of a liquid to produce effects not shown by the bulk liquid. Perhaps the most familiar surface effects are the following: a liquid, such as water, will rise in a capillary tube dipped into the free surface of the liquid; a drop of one liquid deposited on the surface of a second liquid will spread over this surface; mercury can form droplets which become more spherical in shape as their size decreases; small ripples can travel along the surface of a liquid.

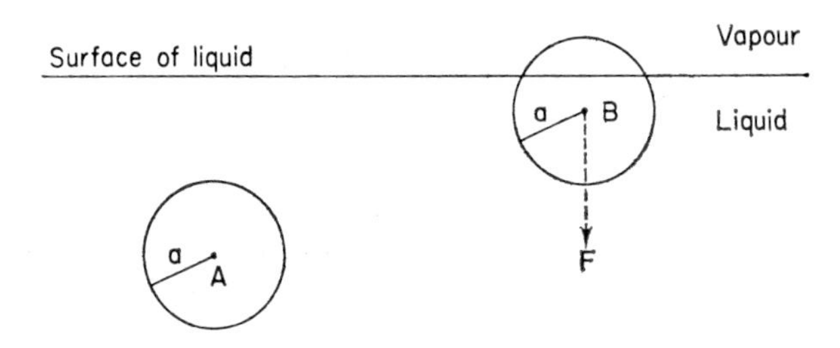

Fig. 7.1. ' Spheres of influence ' around a molecule A in the bulk liquid and around a molecule B in the surface layer.

We start by considering the intermolecular forces which are operating. We refer back to fig. 4.1 which shows the (force, separation) curve for two molecules. When r is very small there is a strong repulsion between the molecules. As r exceeds the equilibrium separation r_0 the force changes to an attraction which tails off to virtually zero at about a distance a which is only a few molecular diameters (see section 4.5).

Let us first consider a molecule A inside the bulk liquid as in fig. 7.1. Draw a sphere of centre A and radius a. From the above discussion the molecules outside this sphere will not exert any attractions on A. The molecules inside this sphere, which number a few hundred at most, will exert forces on A; all these forces will be attractive except for a few possible repulsions produced by A's nearest neighbours. The sphere we have drawn may be termed the ' sphere of influence ' of molecule A. The resultant force on A will vary with time because of the movement of the molecules within this sphere. But the time-average of this force, over a period which is much greater than the period of molecular vibration, will be zero. Since the distance a is of the order of 10^{-9} m then any molecule which is at a greater depth below the surface may be considered to be in the ' bulk ' liquid. For example Tabor quotes some recent experimental work which shows that a water film of thickness 5×10^{-9} m exhibits ' bulk ' properties.

Next consider a molecule B whose distance from the surface is less than a so that a part of its sphere of influence is above the surface in the vapour phase where the concentration of molecules is relatively small. The molecules which are nearest neighbours of B can exert repulsions or attractions on B (depending on their distance from B) but the time-average of these forces will be zero. The remaining molecules of the liquid within the sphere below the surface will exert attractions on B; since there are more of these below B than above it the time-average F of the steady force on B will be downwards and normal to the surface. The nearer B is to the surface the greater the magnitude of F. So molecules such as B whose spheres of influence intersect the surface are said to form an ' anomalous ' surface layer of thickness a. Molecules in this layer are acted on by an inward force and are therefore accelerated inwards from the surface. Thus the number of molecules in the surface, and hence the surface area itself, tends to decrease. When this state of minimum surface area is achieved a dynamic equilibrium is set up in which molecules leaving the surface inwardly are continually replaced by others diffusing upwards from the interior of the liquid.

As a molecule is moved from the bulk liquid to the surface a change in its energy occurs. This may be seen by considering the bonds between the molecule and its nearest neighbours. To break such a bond requires energy so that the energy of a molecule increases as its number of nearest-neighbour bonds decreases. Since a molecule in the actual surface has, on average, half as many bonds as a molecule in the bulk liquid, the surface molecule has the higher energy (see section 7.4).

The surface of a liquid can thus be regarded as the seat of potential energy. Thus contraction of a surface is associated with a decrease of potential energy and vice versa. To enlarge the surface area of a liquid work must be done to bring up more molecules from the bulk liquid. On the macroscopic scale this leads us to define the *free surface energy*, γ, as the work that must be done to increase the surface area by unity under isothermal conditions. Thus γ has the units J m^{-2}.

7.2. *The reality of surface tension forces*

In section 7.1 we introduced the concept of surface energy but this does not of itself give a direct explanation of the well-known tension which exists in a liquid surface. Walton has recently discussed this surface tension and his ideas may be considered by reference to fig. 7.2 in which a section of a liquid film is stretched between two wires A and B. If B is fixed and A free to move then an external force F must be applied to A to maintain equilibrium. If we consider any section C of the film it follows that, for equilibrium, the portion CB of the film must exert on the portion AC a force equal and opposite to F. Does this tension in the film act all the way across the film thickness (as in a stretched sheet of a solid material) or does it act in the surface regions only? That is, is it a bulk tension or a surface tension? It is in fact a surface tension because F remains unchanged as the film thickness is reduced by evaporation. If F is increased in value, and maintained, the film will expand until it breaks. Throughout this expansion the surface tension remains constant even though the film's area is increasing.

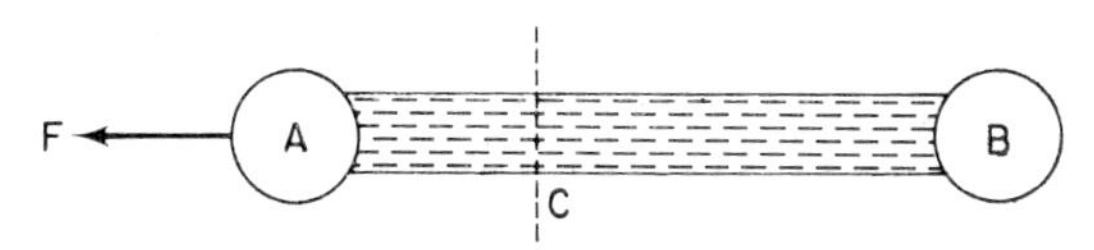

Fig. 7.2. Equilibrium of a liquid film stretched between two wires A and B.

We are now in a position to define surface tension as follows. Consider a short straight line of length Δs drawn in a liquid surface. Then the surface on one side of this line exerts a force Δf upon that on the other side, this force being both tangential to the surface and perpendicular to the line. The surface tension is then defined as

$$\gamma = \frac{\Delta f}{\Delta s} \tag{7.1}$$

and γ, thus defined, has the units N m^{-1}. (Dimensionally the unit of surface energy, J m^{-2}, and that of surface tension, N m^{-1}, are the same.)

The surface tension, as defined in equation (7.1), can be shown to be *numerically* equal to the free surface energy defined in section (7.1). This equivalence, which we shall justify in section 7.3, has already been tacitly assumed by using the same symbol γ to represent both quantities.

A direct explanation of the surface tension in terms of intermolecular forces follows from the fact that the molecules near the surface of a liquid are further apart than those in its interior. There is a good deal of evidence to show that the surface layer is depleted, that is, the number density ρ of the molecules is less than that in the bulk liquid. Experiments using reflected polarized light suggest that this decreased number density is confined to a layer whose thickness is just a few (that is, three or four) molecular diameters.

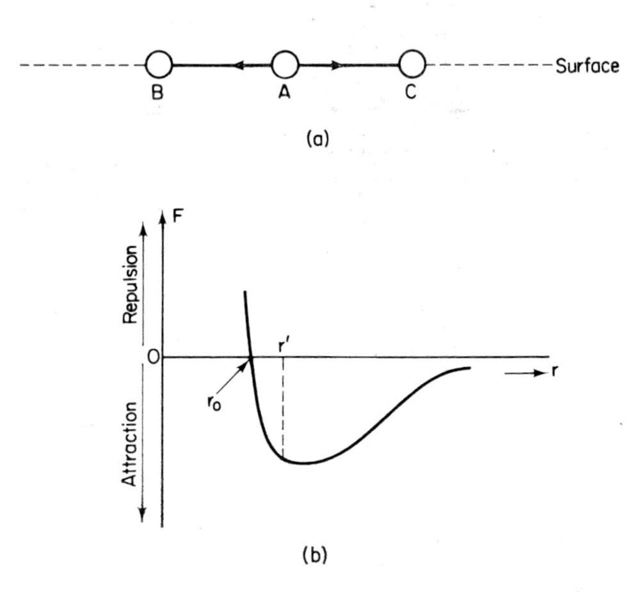

Fig. 7.3. Illustrating the increased molecular separation in the surface of a liquid.

Walton also shows that surface tension occurs as a consequence of assuming that the number of molecules which diffuse from the bulk o a liquid to its surface per second equals the number which leave the surface for the bulk. Because of the depletion in the surface laye the average distance between neighbouring surface molecules is r which is greater than the average ' bulk ' equilibrium separation r discussed in section 4.1. Thus molecule A in fig. 7.3 (a) must nov escape from the *increased* attractions of neighbours such as B and C before it can enter the bulk liquid. These increased attractions occu because the mean intermolecular separation has increased from abou r_0 to r' (fig. 7.3 (b)). These forces, acting parallel to the surface on typical surface molecule like A, therefore give rise to a *surface tensio* according to Walton's theory.

7.3. *Surface tension and surface energy*

Consider a liquid film on a wire frame one of whose sides AB can move without friction (fig. 7.4). Then since the film has two sides a force of $F = 2\gamma l$ must be applied to AB to balance the surface tension forces, where γ is the surface tension defined in equation (7.1). If now the wire is moved slowly, and hence isothermally, through a distance x to a new position A′B′, the work done by the force F is $Fx = 2\gamma lx = \gamma A$, where $A = 2lx$ is the total area of new film surface formed. Hence the work done to increase the surface by unit area is γ. In other words the free surface energy is numerically equal to the surface tension.

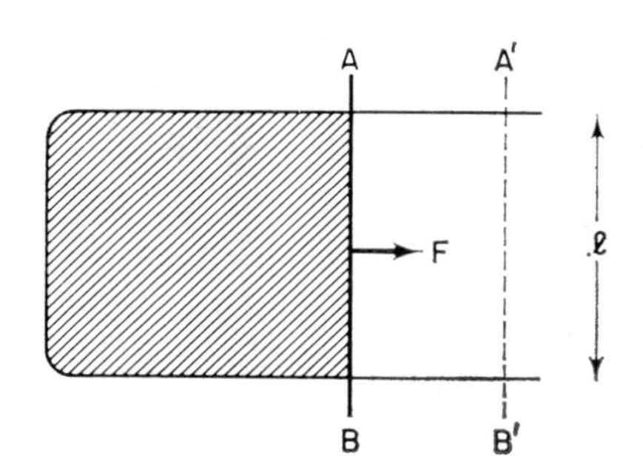

Fig. 7.4. Illustrating the equivalence of free surface energy and surface tension.

When any change of a liquid surface area occurs the *total* change ı energy of the surface is not usually equal to the change in the free ırface energy. This is because γ decreases as the temperature rises. f the surface is increased isothermally the liquid will absorb heat om its surroundings; if it is enlarged adiabatically it will cool. If we ssume that γ is a function of temperature only and does not depend ı the area of the surface, it can be shown that the total energy E quired for the creation of a new unit surface is given by

$$E = \gamma - T\frac{d\gamma}{dT}. \qquad (7.2)$$

his result can be proved thermodynamically (see, for example, *Heat ıd Thermodynamics*, by M. W. Zemansky, 5th edition, Chapter 13). Equation (7.2) shows that when a surface area increases isothermally e total surface energy E is made up of two parts, namely (*a*) the free rface energy γ due to the enlarging of the surface and (*b*) the ‘ latent at ’ term $-T(d\gamma/dT)$ which must be provided to keep the temrature constant. (Since $d\gamma/dT$ is negative this ‘ latent heat ’ term actually positive.) For water in contact with its vapour at room nperature, $\gamma = 72 \times 10^{-3}$ J m^{-2} and $E = 118 \times 10^{-3}$ J m^{-2}.

We can interpret equation (7.2) in terms of our previous molecular model of surface tension. When a surface is extended, some extra molecules have to be brought from the interior to the new surface, and the average separation of these particular molecules is thus increased. This means that work has to be done against the attractive forces between these molecules. The work done against the surface tension provides a part of this work equal in amount to γ J per square metre of surface. But the remainder of this work is provided by the

Fig. 7.5. A drop of water formed in anisole (after J. Cunliffe).

vibrational energy of the molecules, which therefore decreases. It follows that an equivalent amount of heat will flow into the surface from the surroundings, and this is represented by the second term on the right-hand side of equation (7.2).

It must be emphasized that the value of $\gamma = 72 \times 10^{-3}$ J m^{-2} for water quoted above is the value when the water is in contact with its vapour, that is, it is the value for a liquid-vapour interface (the effect of the atmosphere being negligible). By contrast the value of γ for a water–chloroform interface at room temperature is about 28×10^{-3} J m^{-2}. All this emphasizes the fact that when the second medium is not the liquid vapour but is another liquid (or solid even) this fact must be clearly stated. Fig. 7.5 shows the interface between two liquids where a drop of water is formed at the end of a tube immersed in a beaker of anisole.

7.4. *Surface energy and nearest-neighbour bonds*

We now obtain an estimate for the value of the surface energy in terms of the molecular diameter σ and the depth ϵ of the mutual potential energy curve of fig. 4.2.

A molecule in the bulk liquid is surrounded on average by z nearest neighbours each at a distance of about r_0 from it. In the surface each molecule has $\frac{1}{2}z$ nearest neighbours. Thus to create a new liquid surface some nearest-neighbour bonds will have to be cut. Suppose herefore that a liquid column of unit cross-sectional area is 'sliced' across so as to produce two new surfaces S_1 and S_2 each of unit area fig. 7.6).

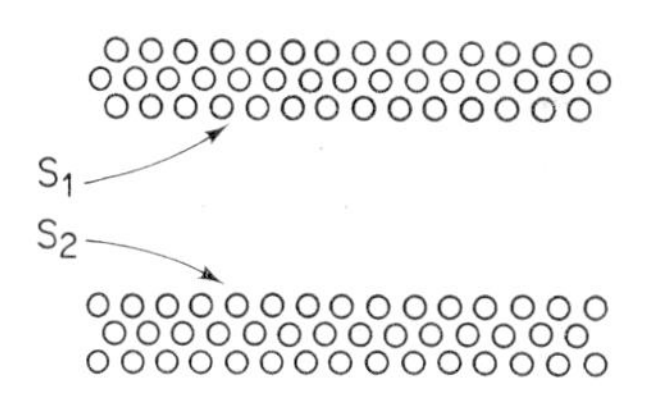

Fig. 7.6. Creating two new unit surfaces S_1 and S_2 by 'slicing' a liquid column in two.

Let there be n molecules in a cross-section of the liquid column. fter the separation of the two parts of the column each molecule in oth surfaces will have $\frac{1}{2}z$ nearest neighbours and not z as before the reak. Thus in producing the two new unit surfaces the number of earest-neighbour bonds that have been cut is $\frac{1}{2}nz$. Since an energy ϵ is required to cut each bond (see section 4.5) the total energy quired is $\frac{1}{2}nz\epsilon$. Hence the energy needed to produce a *single* new unit area of surface is $\frac{1}{4}nz\epsilon$ which we equate to the surface energy. It is easy to show that n is also about $1/\sigma^2$ and so $z\epsilon/4\sigma^2$ is a good estimate for the surface energy. In all this it is assumed that the surfaces S_1 and S_2 are created isothermally and reversibly.

For liquid argon the relevant data are

$$\epsilon = 1{\cdot}653 \times 10^{-21}\ \text{J}$$

$$\sigma = 3{\cdot}405 \times 10^{-10}\ \text{m}.$$

If we take $z = 10$, this gives us

$$\frac{z\epsilon}{4\sigma^2} = 3{\cdot}567 \times 10^{-2}\ \text{J m}^{-2} = 35{\cdot}67\ \text{mN m}^{-1}$$

which is close to the experimental value given by Stansfield (Stansfield, D., 1958, *Proc. Phys. Soc.*, **72,** 854).

7.5. *Theories of surface tension*

(*a*) *Berry's theory*

This is essentially a hydrostatic theory to explain how a tension parallel to the liquid surface arises. If we consider a small imaginary area (a ' test surface ') at a point inside a fluid in equilibrium then the pressure at that point may be defined as the average normal force per unit area exerted by all the molecules on one side of the surface on all those on the other. Berry regards this pressure p as being made up of two contributions. One part p_K is due to the transport of momentum by molecules (as in a gas) and its value is $p_K = \rho kT$ where ρ is the number density of molecules at the region in question p_K is always positive. The second contribution p_f is due to the time average of the interactions between molecules on opposite sides o the test surface (cf. the term a/V^2 in van der Waals' equation) and i particularly relevant when discussing a liquid or dense gas. We the write

$$p = p_K + p_f.$$

p_K is much greater for a liquid than for a gas. If the externall applied pressure is not too large the molecules are not so very closel packed and so the attractive forces between the molecules dominat the repulsive forces; in this case p_f is negative in order to reduce th total pressure p to the value of that applied externally. If the extern pressure gets very high the molecules are squashed together and th repulsive forces predominate and this further compression is resisted p_f is then positive.

Let us refer to fig. 7.7. In the bulk vapour (A) or bulk liquid (C there is directional symmetry in the distribution of the molecules an so the pressure has the same value p_0 irrespective of the orientatio of the test surface used to define it. However, in the surface region (B), the tangential pressure p_t and the normal pressure p_n will not be the same because there is no longer symmetry of direction (there is depletion in the vertical z direction but not in the horizontal x direction). Since, however, the fluid is in equilibrium, the forces on the opposite faces of the three small cubes shown must be equal and opposite. Thus, neglecting gravity changes between A and C, the normal pressure p_n has the same constant value p_0 between A and C, that is, right through the surface layer. On the other hand, the tangential pressure p_t is equal to p_0 in the bulk vapour at A and in the bulk liquid at C but is *not* equal to p_0 in the surface layer at B.

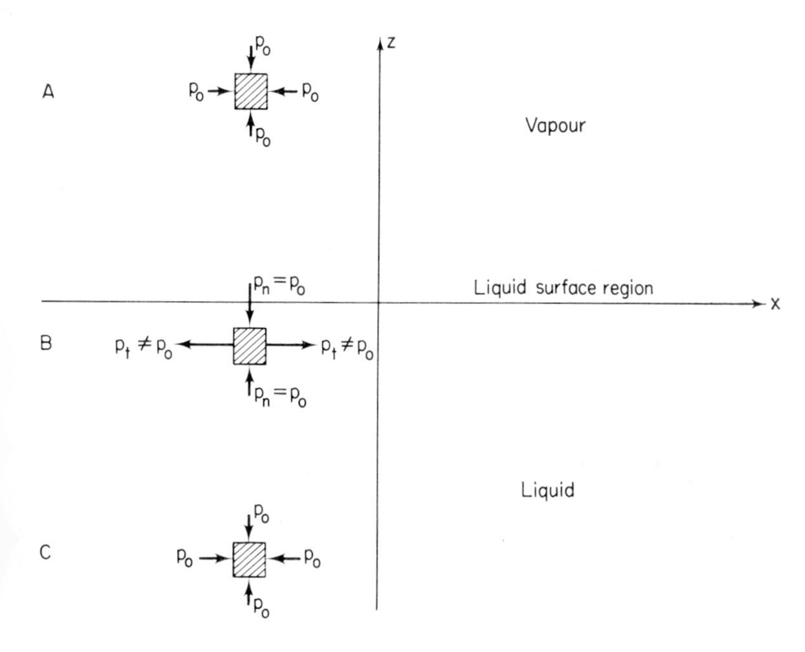

Fig. 7.7. Equilibrium of elementary fluid cubes in liquid, vapour and surface, illustrating Berry's theory.

Berry then proceeds to explain how the two contributions p_{kt}, p_{ft} o p_t vary with z. We shall not give the argument here but refer the nterested reader to the original paper. Berry concludes that p_t 'aries with z in the manner shown by the dotted curve in fig. 7.8. As we see this curve goes negative showing that there is a tangential ension near the liquid surface. Berry concludes that " the surface ayer of liquid, in contrast with the bulk, must possess rigidity in order to resist the shear stress that results from p_t differing from p_n; this is the basis for the statement appearing in older textbooks that liquids behave as if their surfaces are covered by an ' elastic skin ' ".

Ideas similar to those of Berry have previously been considered by Bakker and by Brown (see references).

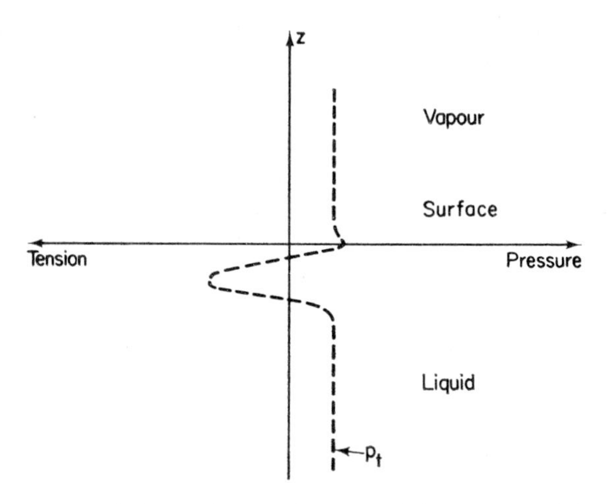

Fig. 7.8. The variation with z of the tangential pressure p_t (after Berry).

(b) Walton's theory

We have already partly discussed this theory in section 7.2 where we described how Walton used the ideas of molecular diffusion to explain the existence of surface tension. In this treatment he uses the cell model of a liquid in which each molecule is surrounded by about 10 nearest neighbours. He argues that the average potential energy ϕ' of an enclosed molecule as it is displaced along a chosen direction x is as shown in fig. 7.9. At A, the molecule is at its cell

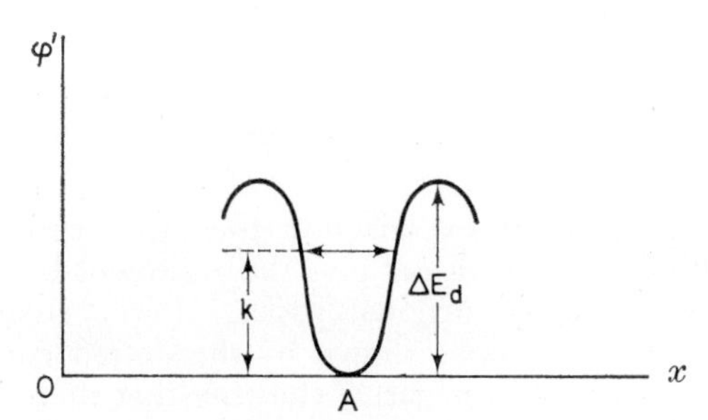

Fig. 7.9. The average potential energy ϕ' of a molecule in a cell as it is displaced in the x direction.

centre. To escape from the cell to the right in the x direction it must acquire a total energy ΔE_{d}, the diffusion activation energy. When the kinetic energy K of the enclosed molecule exceeds ΔE_{d}, escape is possible and diffusive flow occurs. This is however a slow process; for most of the time the molecule is confined to its cell and only occasionally will K be large enough for it to escape.

Walton also uses similar diffusion arguments to explain capillarity. He considers how a liquid rises in a glass capillary tube which is lowered into the liquid. He assumes that a molecule of the liquid first requires an energy ΔE_{e} to escape upwards from the liquid (the latent heat of vaporization). If this molecule then settles on the glass wall of the capillary an amount of energy ΔE_{a} will be surrendered (the heat of absorption). Thus to escape from the liquid to the glass a molecule must have a kinetic energy of at least ΔE_{e}; conversely a molecule which has settled in a site on the glass requires a kinetic energy of at least ΔE_{a} to return to the bulk liquid. If $\Delta E_{\mathrm{a}} > \Delta E_{\mathrm{e}}$ more molecules will diffuse up the glass wall from the bulk liquid than vice versa, that is, *capillary rise* occurs. As a molecule A at the top perimeter of the meniscus diffuses up the glass wall it pulls on other neighbouring molecules; this explains the upward force around the perimeter of the tube at the top of the meniscus. As the rise of the liquid in the tube continues, further diffusion up the glass wall becomes increasingly difficult. Molecules like A at the wall at the top of the meniscus now experience the downward weight of the liquid column; this means that an energy greater than the initial ΔE_{e} is now required to escape from the bulk liquid. Eventually a state of dynamic equilibrium between those molecules diffusing up the glass from the liquid and those returning down the glass to the liquid will be reached when the weight contribution on A is sufficient to increase the value of ΔE_{e} to ΔE_{a}; that is, no further capillary rise occurs.

All the above discussion emphasizes the fact that capillarity does not depend on the liquid alone but also on the nature of the wall of the capillary tube, including its cleanliness. For example, it has been shown that the extent of the capillary rise in a glass capillary depends on the ionic content of the glass. Again, depending on the degree of cleanliness present, mercury can sometimes rise in a capillary and sometimes (most often) be depressed.

(c) *Other theories*

There are other, more sophisticated, theories of surface tension involving the use of statistical mechanics and requiring a knowledge of such quantities as the radial distribution function, intermolecular potential energy, etc. Such theories are too difficult for us to consider here but details may be found in a paper by R. C. Brown.

7.6. *Interfaces and Young's relation*

We now concentrate on the surface *energy* approach and consider a liquid in contact with a plane solid surface as in fig. 7.10 (*a*). There are three phases involved—solid, liquid and vapour and hence three interfaces—solid–liquid, liquid–vapour and solid–vapour; to these interfaces there will be free surface energies of γ_{SL}, γ_{LV} and γ_{SV} respectively. In the diagram an angle θ, called the angle of contact, is shown. This is the angle between the tangential planes to the solid–liquid and liquid–vapour interfaces at the line of contact with the solid and it is measured in the liquid as shown. We now derive a relation between the three free surface energies and θ.

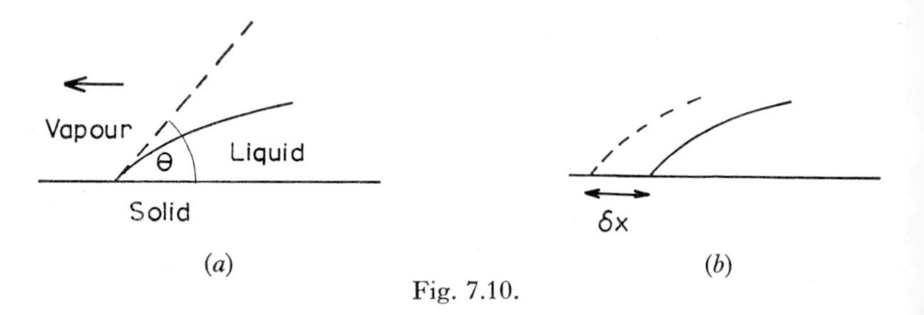

Fig. 7.10.

For this purpose we suppose that the liquid in fig. 7.10 (*a*) extends to infinity over the plane surface of the solid to the right of the line of contact. Consider now a movement of the liquid to the left in the direction of the arrow in fig. 7.10 (*a*) so that the line of contact moves through a small distance δx as in fig. 7.10 (*b*); then the overall potential energy will not change. Three area changes will be involved if we consider unit length of the line of contact perpendicular to the plane of the diagram, namely (see fig. 7.11):

Area increase of solid–liquid interface $= \delta x$

Area decrease of solid–vapour interface $= \delta x$

Area increase at liquid–vapour interface $= \delta x \cos \theta$.

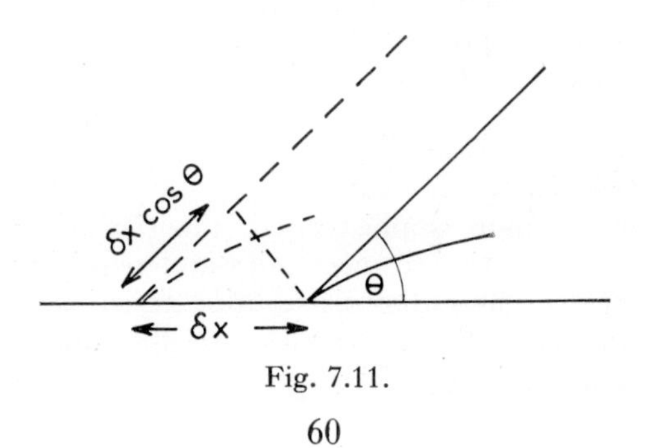

Fig. 7.11.

By the principle of virtual work the total change in potential energy is zero and so

$$\gamma_{SL}\,\delta x+\gamma_{LV}\,\delta x\cos\theta-\gamma_{SV}\,\delta x=0$$

or

$$\gamma_{SV}=\gamma_{SL}+\gamma_{LV}\cos\theta. \tag{7.3}$$

This is the general relation between the three interfacial free surface energies. It is known as Young's relation after T. Young who gave it in 1805 and it was later derived by A. Dupré in 1869.

In a situation like that shown in fig. 7.10 (*a*) the angle θ will depend on the nature of the solid and the liquid (and to a much less extent on the vapour) or, more specifically, on the relative magnitudes of the forces exerted on a molecule of the liquid by the liquid molecules and the solid molecules in its immediate neighbourhood. We can distinguish between three cases:

(*a*) $0^\circ<\theta<90^\circ$. An example is water in contact with a slightly greasy glass surface. This is referred to as 'imperfect wetting' or 'partially wetting'.

(*b*) $\theta=0^\circ$. In certain cases where there is a strong attraction between the liquid and solid molecules the angle of contact is very small or zero, an example being very clean glass and water; in this case the liquid is said to 'wet' the solid.

(*c*) $\theta>90^\circ$. Now we say that the liquid 'does not wet' the solid, the best example being mercury and clean glass where θ may be about 136°.

There are various ways of measuring the 'wettability' of solids by liquids which are, in essence, the same as specifying the angle of contact. This angle is not always a constant but depends on whether the liquid is advancing over the solid surface or receding. These two angles are often known as advancing and receding angles of contact and θ_A is usually larger than θ_R (fig. 7.12).

Advancing　　　　Receding

Fig. 7.12.

7.7. *Spreading*

Consider two liquids A and B whose free energies are respectively γ_A and γ_B (these are the values of γ in contact with the vapour phases). If we take a column of liquid A of unit cross-section and assume that this column can be separated across a plane cross-section as in fig. 7.13 then two new surfaces of unit area are created. Thus there is an increase of free surface energy of $2\gamma_A$ and $W_A = 2\gamma_A$ is called the *work of cohesion* of liquid A. In the same way $W_B = 2\gamma_B$ is defined similarly for liquid B.

Suppose further that the two liquids A and B do not mix and consider a column of unit cross-section of the two liquids in contact as in fig. 7.14 (*a*). Liquid A rests on liquid B and there is a plane interface between them. If we assume that the two liquids can be separated at this interface the work done equals the change in free surface energy. This is $W_{AB} = \gamma_A + \gamma_B - \gamma_{AB}$, since two new unit surfaces of energies γ_A, γ_B have been produced where previously there was an AB interface. W_{AB} is called the *work of adhesion*.

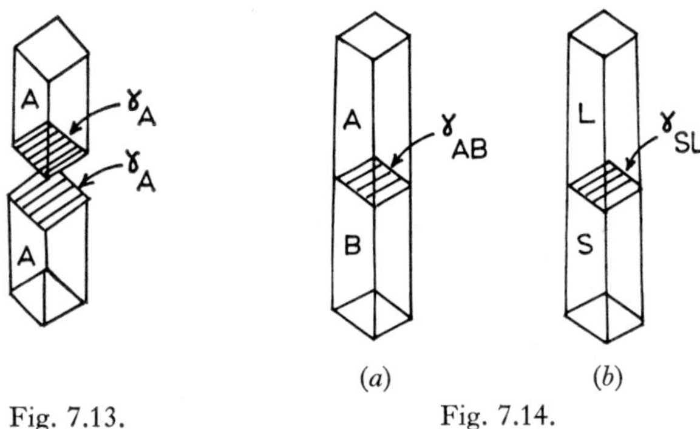

Fig. 7.13. Fig. 7.14.

Similarly if a liquid L is in contact with a plane solid surface S then $W_{SL} = \gamma_S + \gamma_L - \gamma_{SL}$ is the work of adhesion (fig. 7.14 (*b*)).

Values of the work of adhesion for various organic substances in contact with water are given in table 7.1 below; they are taken from a Unilever Educational Booklet *Surface Activity*, page 7.

Substance	$\gamma_{AB} \times 10^3$ (J m^{-2})
Paraffins	36–48
Aromatic hydrocarbons	63–67
Alkyl halides	66–81
Esters	73–78
Ketones	85–90
Nitriles	85–90
Primary alcohols	92–97
Fatty acids	90–100

Table 7.1.

Now consider the problem of a liquid spreading (*a*) on a plane horizontal solid surface and (*b*) on another liquid with which it does not mix.

Take first the surface of a solid. The reason why such a surface has a free surface energy γ_{SV} is because the molecules in it do not have a symmetrical environment; on one side of the surface there are neighbouring molecules of the solid, on the other only a few vapour molecules. If now this solid surface is covered by a liquid then this largely redresses the imbalance in the environment of the surface molecules—there are now comparable concentrations of neighbouring molecules both above and below the actual surface—and so γ_{SV} is reduced to γ_{SL}.

If the liquid spreads over the solid's surface the liquid–vapour interface increases and so does the liquid-solid interface whilst the solid–vapour one decreases. These changes cause the various free surface energies to change and spreading will proceed until the sum of the free energies is a minimum, that is, provided that the increase in the free surface energy accompanying the area increase in the LV and LS interfaces is less than the decrease in energy due to the area decrease in the SV interface, that is, if

$$\gamma_{LV} + \gamma_{SL} < \gamma_{SV}.$$

It is assumed in all this that the spreading film is thick enough for us to regard γ_{SL} and γ_{LV} as being independent; in a film which was only a few molecules in thickness one would have to look into this assumption in more detail.

Next consider two immiscible liquids A and B. If a drop of A is introduced on to the plane surface of B then the argument is the same as that above for spreading over a solid surface; spreading now occurs if

$$\gamma_A + \gamma_{AB} < \gamma_B. \qquad (7.4)$$

To sum up, if spreading of a liquid occurs either over a solid or a second liquid surface the spreading liquid moves so as to reduce the sum of the free surface energies.

If we return to the case of two liquids we can define a spreading coefficient S as

$$S = \gamma_B - \gamma_A - \gamma_{AB}.$$

Hence, from (7.4), spreading occurs if $S > 0$. Also, since

$$W_{AB} = \gamma_A + \gamma_B - \gamma_{AB}$$

and

$$W_A = 2\gamma_A$$

then

$$S = W_{AB} - W_A.$$

Thus spreading occurs if $W_{AB} > W_A$, that is, if the work of adhesion for the two liquids is greater than the work of cohesion for the spreading liquid.

The equations

$$S = \gamma_B - \gamma_A - \gamma_{AB}$$

and

$$S = \gamma_{SV} - \gamma_{LV} - \gamma_{SL}$$

really express the spreading behaviour of a so-called ' duplex ' film. Such a film is of sufficient thickness for γ_A and γ_{AB} (or γ_{LV} and γ_{SL}) to act independently, that is, the γ's for the upper and lower surfaces of the spreading liquid are independent. Duplex films are not thermodynamically stable and usually change into monomolecular films (monolayers). Often indeed after the initial spreading as a duplex film the molecules in the spreading liquid will rearrange themselves into liquid ' lenses ' in equilibrium with the monolayer. An example of this is benzene on water.

A monolayer behaves quite differently from both the substrate below it and the bulk of the liquid from which it is composed. In the spreading of monolayers the minimizing of the total free surface energy is again the criterion.

References

(1) Noakes, G. R., *New Intermediate Physics*, Chapter 9 (Macmillan & Co. Ltd., London, 1957).

(2) Brown, R. C., 1974, *Contemporary Physics*, **15**, 301.

(3) Berry, M. V., 1971, *Physics Education*, **6**, 79.

(4) Walton, A. J., 1972, *Physics Education*, **7**, 491.

(5) Zemansky, M. W., *Heat and Thermodynamics*, 5th Edition, Chapter 13 (McGraw-Hill Book Co., 1968).

(6) Sprackling, M. T., 1970, *The Mechanical Properties of Matter*, Chapter 7 (The English Universities Press Ltd., London).

CHAPTER 8

stretching and compressing a liquid

Part 1. STRETCHING A LIQUID

8.1. *Introduction*

We begin by asking a question: is it possible to stretch a liquid? Put more scientifically, is a liquid capable of withstanding tension? Rather surprisingly the answer is ‘ yes ’ and this property of liquids is of great interest in both pure and applied science. In physics it sheds some light on the equation of state of a liquid. In marine engineering, concerned with designing such things as ship propellers, the pressure in the water near the propeller blades often falls to a negative value, that is, tension is set up. If this tension is sufficient the liquid ‘ breaks ’ and this ‘ break ’ takes the form of ‘ cavitation ’, the name given to the production of a number of very small cavities in the liquid. Cavitation can also occur when a liquid flows rapidly in a pipe; near a constriction in the pipe the velocity becomes very high and the pressure can drop to zero or negative values producing the vibration known as ‘ hammering ’. This phenomenon of cavitation has been very widely studied and the value of the tension necessary to cause its onset is often called the ‘ critical tension ’ or ‘ cavitation threshold ’ of the liquid (see Frontispiece).

In botany, we know that columns of sap transport water from the roots of a tree to the uppermost leaves. Now the atmosphere will support a column of water only 10·4 m high and since some trees are far taller than this then the only way in which the sap can be drawn up to these greater heights is by the existence of a negative pressure in the sap column. Some recent work has shown that negative pressures of -50 atm exist in mangrove trees; even higher negative pressures of up to -80 atm have been measured in desert plants n their efforts to suck up every drop of water from their dry environment.

The ability of liquids to withstand tension was first predicted by Laplace in his theory of capillarity and since then a good deal of xperimental and theoretical work has been done in this field. This work will now be described.

8.2. *The behaviour of liquids under hydrostatic tension*

The pioneer experiments in this field were those reported by Berthelot in 1850. His method is now usually known as the Berthelot tube method ’. A Berthelot tube is a sealed cylindrical tube which, at ordinary room temperatures, is almost completely filled with a liquid while the remaining volume is occupied by air and the liquid vapour. When the tube and its contents are heated the liquid expands at a greater rate than the Berthelot tube, the air is forced into solution and the liquid fills the tube completely at a certain temperature T_f. If the tube is then cooled it is found that the liquid adheres to the whole of the inside wall of the tube and continues to fill it at temperatures below T_f. So as the temperature falls below T_f a gradually increasing uniform tension F is built up in the liquid until it eventually ruptures at a lower temperature T_b. This instant of rupture is invariably accompanied by an audible metallic ' click ' and also a sudden increase ΔV in the external volume of the Berthelot tube due to the release of tension. The various stages in a Berthelot tube experiment are shown in fig. 8.1.

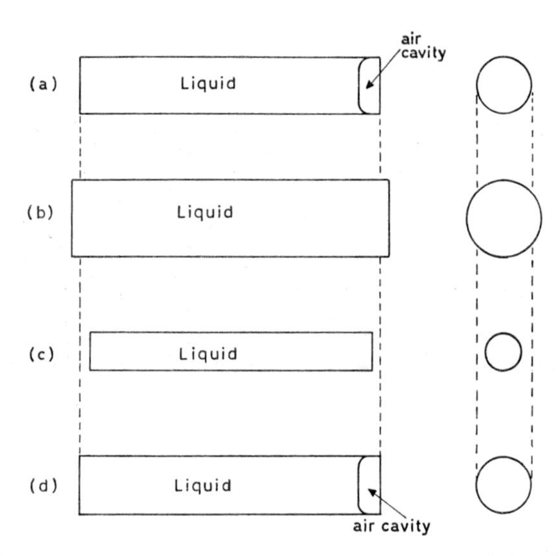

Fig. 8.1. (*a*) Berthelot tube, at room temperature, containing liquid and a small volume of air and liquid vapour. (*b*) Tube after being heated from room temperature until the liquid just fills it at the temperature T_f. (*c*) Tube after being cooled to a temperature T_b (below T_f) and just prior to rupture. Between stages (*b*) and (*c*) the liquid is subjected to a gradually increasing tension. (*d*) Tube after the liquid has fractured

One way of estimating the critical tension $\hat{F}$ in the liquid just before it breaks is to use an equation which relates $\hat{F}$ to ΔV and the dimensions and elastic constants of the tube. Temperley and Chambers used this method to measure the critical tension of water in glass Berthelot tubes. The value of ΔV was estimated by enclosing the tube in a glass jacket connected to a vertical capillary; water occupied the remainder of this glass jacket and also filled part of the capillary. The change in the position of the meniscus in the capillary at the time of rupture then enabled ΔV to be obtained. The results showed that the critical tension which water could sustain in a glass tube was between 30 and 50 atm. Rees and Trevena repeated these experiments for water in steel Berthelot tubes and found that the critical tension was consistently less and varied between 10 and 30 atm. Temperley further found that if he placed a small chip of steel inside his glass tube the critical tension was reduced to values similar to those obtained in the steel tubes. All this strongly suggests that what is really measured by a Berthelot tube is the adhesion of the liquid to the surface rather than its true tensile strength. The measured value of the tension is that of the ' weakest link ' in the liquid–solid system and in the experiments mentioned above the water–steel link proved to be weaker than the water–glass one.

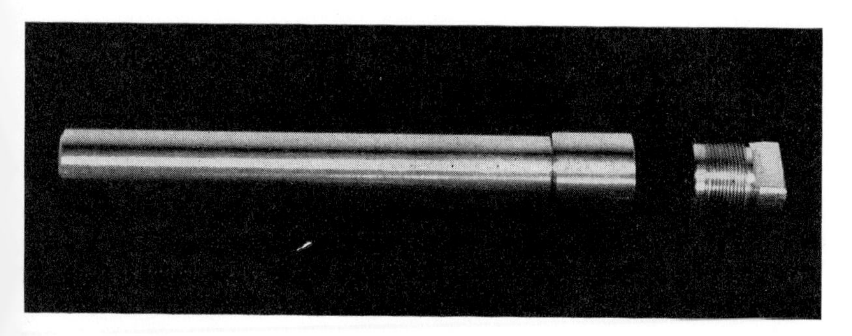

Fig. 8.2. Photograph of a steel Berthelot tube used by author and his co-workers. It is closed at one end and is then sealed at the other by means of a threaded cap, shown removed.

More elaborate forms of Berthelot tube have been developed in recent years. In one type of stainless steel tube Hayward measured the negative pressures inside by fitting strain gauges to its outside wall. Still more recently Richards has used a stainless steel tube with a strain-gauge pressure transducer fitted so that the diaphragm of the transducer actually formed one end of the tube. In this way the variations of pressure and tension in the enclosed liquid could be followed directly over various temperature ranges.

A centrifugal method of stretching a liquid was devised by Osborne Reynolds in 1878. His liquid was contained in a glass J-tube in which the long arm was sealed (so that the liquid was simply under its own vapour pressure) or left open to the atmosphere. The latter type of tube is shown in fig. 8.3. If such a tube is rotated with angular velocity ω about an axis O perpendicular to the horizontal plane containing the tube axis, a pressure gradient of $dp/dr = \rho\omega^2 r$ is generated in the direction OA, where r is the distance from O. Since the pressure p_A at A is always atmospheric it follows that the pressure p_O at O is less than atmospheric, and we have

$$p_A - p_O = \int_0^R \rho\omega^2 r \, dr = \tfrac{1}{2}\rho\omega^2 R^2$$

where

$$OA = R.$$

Thus

$$p_O = p_A - \tfrac{1}{2}\rho\omega^2 R^2.$$

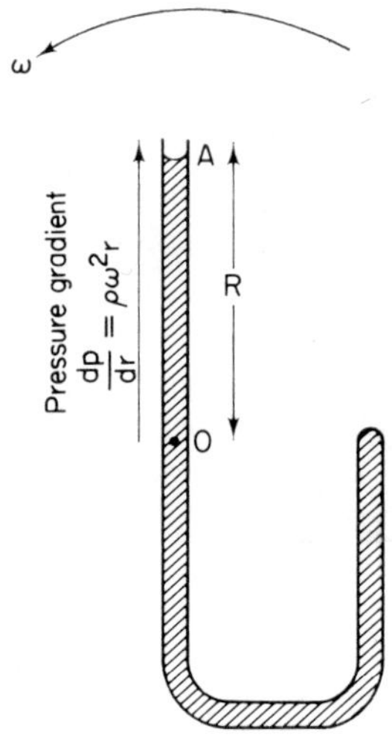

Fig. 8.3. J-tube used by Osborne Reynolds rotating in a horizontal plane about a vertical axis through O.

So if ω is sufficiently large p_O becomes *negative*, that is, tension sets in and when ω reaches a certain value ω_c the liquid ruptures at O. Knowing ω_c, the value of the critical tension can be calculated.

In 1950 L. J. Briggs, who was then head of the National Bureau of Standards, rotated a Z-shaped capillary, open at both ends, filled with a liquid. Briggs emphasizes the need for scrupulous cleanliness in such measurements. In the centrifugal field one-half of the Z-shaped liquid column was pulling against the other and eventually, when ω was high enough, the column broke at its centre which coincided with the axis of rotation. Briggs's value for the critical tension of distilled water in a Pyrex glass capillary is about 277 atm at 10°C—an ' all-time high ' experimental value. He also studied the effect of temperature on this critical tension. At 10°C the value reached a maximum of 277 atm and this fell to 217 atm at 50°C; between 5° and 0°C the value fell dramatically to less than 10 per cent of the maximum value and this represents another anomaly in the behaviour of water in this temperature range. Briggs's results are shown in fig. 8.4.

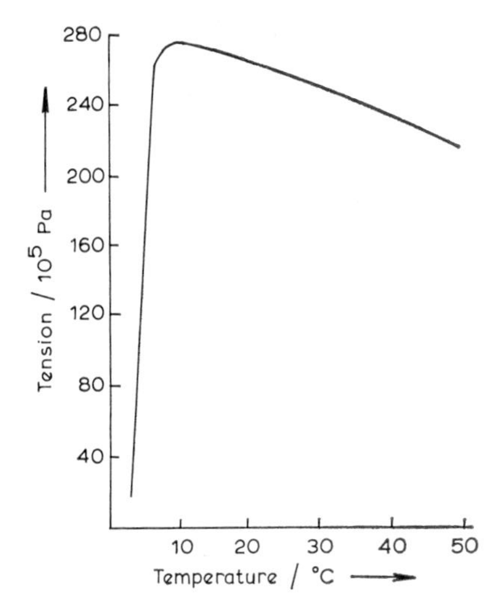

Fig. 8.4. L. J. Briggs's results for water in a Pyrex glass capillary.

:.3. *Dynamic stressing of a liquid*

In the ' static ' experiments described in section 8.2 a gradually ncreasing tension was applied for a comparatively long time so that here was a very low rate of stressing. We now consider the case of dynamic ' stressing in which the tension is applied to a liquid for mes of the order of a millisecond or so. One way of doing this is) have the liquid contained in a vertical cylindrical tube fitted with a eel piston at its lower end. A pressure pulse can then be generated 1 the liquid by firing a lead bullet to strike the lower end of the piston ormally at its centre. This pulse is shown as pulse (1) in fig. 8.5; ie pressure rises rapidly in about 50 μs to a maximum value p and ıen decays more gradually over a few hundred microseconds. For a ulse in water p could be about 300 atm and the total duration τ of ıe pulse was typically 500 μs. When this compressional pulse eaches the upper free surface of the liquid it is reflected as a pulse of nsion. This is because the total pressure at the free surface of the incident and reflected pulses is nearly equal to the atmospheric pressure. Put in another way it is because the acoustic impedance of air is less than that of the liquid. The phenomenon is similar to optical reflection at a rarer medium.

At a depth D below the free surface we then have the initial pressure pulse (1) followed by a tension pulse (2) as shown in fig. 8.5. The time lag between the two pulses is $2D/c$ where c is the velocity of propagation of the pulses in the liquid. In fig. 8.5 we have shown the case $2D/c > \tau$ in which the two pulses do *not* overlap in time and the maximum tension that can be expected in the liquid is F, provided the liquid does not cavitate at a lower value of the tension.

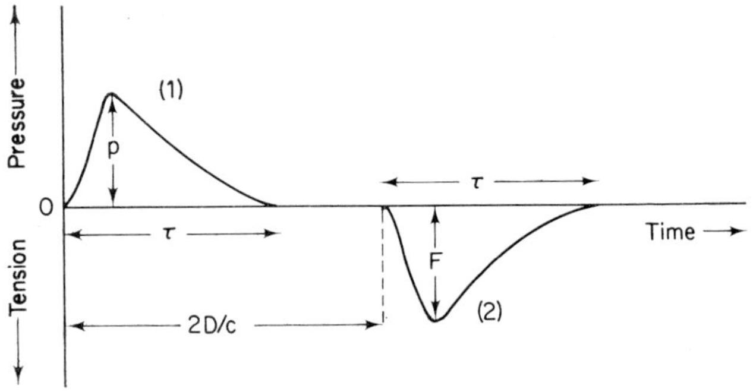

Fig. 8.5. The incident and reflected pulses at a depth D below the free surface of the liquid.

Couzens and Trevena used this method to study the effects at depth D in a liquid and they recorded the pressure changes by means of piezo-electric pressure transducer (fig. 8.6). Many records were take in which p was gradually increased by using different kinds of bulle and pistons of varying masses; for each record the values of p and were noted. The (F,p) curve was then drawn and in each case th curve was similar to that shown in fig. 8.7. This shows that th value of F does not increase linearly with p, as would be expecte from our knowledge of reflection at a free surface, but levels off at constant limiting value $\hat{F}$. This limiting value clearly represents tl maximum tension the liquid can stand and hence is the critical tensic needed to produce cavitation. The critical tensions obtained we $\hat{F} = 8{\cdot}5$ atm for ordinary tap water and $\hat{F} = 15{\cdot}0$ atm for boil deionized water. The higher value of $\hat{F}$ in the second case is wh one would expect since the boiling would remove most of tl dissolved gas; thus there would be fewer cavitation 'nuclei' the form of microscopic gas bubbles present and this would increa the ability of the liquid to withstand tension. Stephen Sedgewick is at present continuing these investigations.

One other important feature of the results obtained from dynamic stressing of liquids must be mentioned. This is that the values of the critical tension obtained under these rapid rates of stressing are consistently lower than those obtained by hydrostatic methods. The explanation for this is not known.

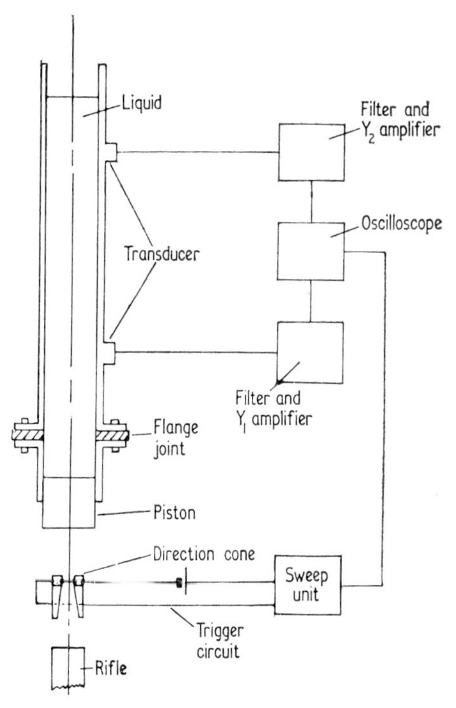

Fig. 8.6. Schematic diagram of the apparatus used by Couzens and Trevena.

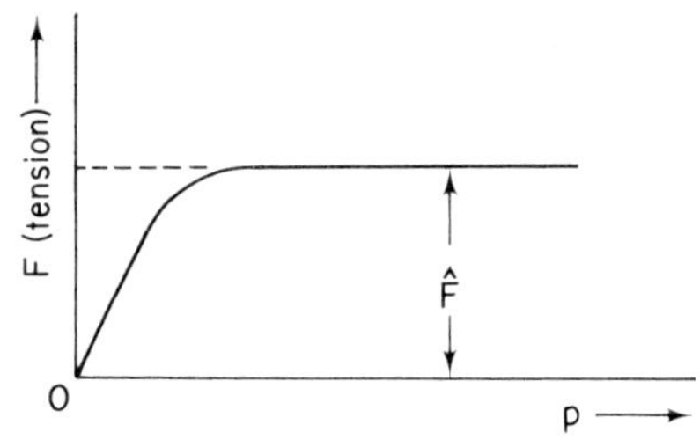

Fig. 8.7. Graph showing how the tension F levels off to the value $\hat{F}$ as the peak pressure p in the incident pulse is increased.

8.4. *Ultrasonic stressing of a liquid*

This type of stressing accompanies the propagation of ultrasonic waves in a liquid. This high-frequency sound wave alternately subjects any point in the liquid to a compression and a tension during the positive and negative half-periods of the pressure cycle. As the tension increases from zero in the negative half-period a stage may be reached when cavitation sets in; as the pressure subsequently increases in the positive half-period the cavitation bubbles contract rapidly.

Ultrasonic stressing has shed much light on cavitation phenomena. There appear to be two main types of cavitation bubble. The first is the gas-filled bubble containing gas (or air) previously dissolved in the liquid and such bubbles usually grow to visible size. The stage at which these bubbles appear is often referred to as ' degassing '. The second type, the vapour-filled bubble, contains the vapour of the surrounding liquid; these bubbles are much smaller than the gas-filled type and their growth and collapse are quite explosive in nature. There is also evidence of a third type of bubble which is a completely empty void.

When a liquid contains dissolved air it cavitates ultrasonically at lower values of the tension than when the liquid is degassed. This difference also occurs, as we saw in section 8.3, when we have dynamic stressing. We are still however far from an understanding of how cavitation is actually initiated but we do know that it can start either at the wall of the containing vessel or in the body of the actual liquid itself.

8.5. *Tensile strength of a van der Waals liquid*

Temperley has shown how to obtain a theoretical estimate of th tensile strength of a liquid which obeys van der Waals' equation Consider two van der Waals isotherms corresponding to tw temperatures T_1 and T_2 as in fig. 8.8.

As we saw earlier, the positions of the liquid–vapour equilibriur lines AB and CD are given by the rule of equal areas (see section 3.4 Along the portions AE, CF of the curves the liquid is in a metastabl phase and at the lower temperature T_2 the part C′F of th metastable phase extends into the negative pressure region. In othe words as we go from C′ to F the liquid is withstanding an increasin tension. Temperley then argues that the point F which represen the limit of the metastable region at this temperature must als represent the limiting value of the tension that the liquid can sustai Thus the ordinate OF′ represents the *tensile strength* of the liqui

Using suitable values for the van der Waals constants a and b an putting T_2 equal to room temperature the value of the tensile streng of water turns out to be about 500 atm. This theoretical value

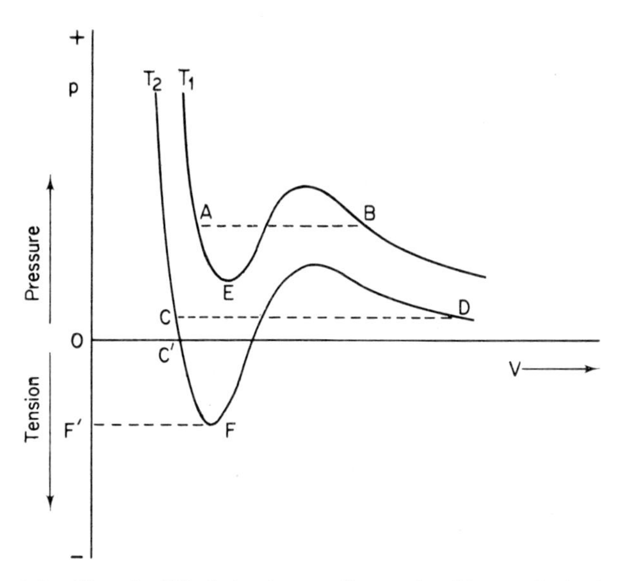

Fig. 8.8. Van der Waals isotherms illustrating Temperley's theory.

500 atm represents the ' true ' tensile strength of water and is much greater than all the *experimental* values obtained for the critical tension (with the exception of L. J. Briggs's centrifugal value of nearly 300 atm discussed in section 8.2). But as we saw earlier the experimental values are a measure of the weakest link in the system and not the true tensile strength, so that the discrepancy between Temperley's theoretical value of 500 atm and the much lower experimental values need not worry us too much.

8.6. *Discussion*

We summarize the position as follows. The first important point is that most of the experimental values reported for the critical tension of a liquid are a measure of the ' weakest link ' in the liquid–container system and are not the values of the tensile strength of the liquid. Furthermore, the experimental values of the critical tension under dynamic stressing are consistently lower than those obtained under hydrostatic conditions. The role of gas and vapour nuclei is also an important factor. In general the theoretical values for the tensile strength of a liquid are much higher than the experimental values obtained for the critical tension.

References

(1) ' The Behaviour of Liquids under Tension '. Article by D. H. TREVENA in *Contemporary Physics*, **8,** 185 (1967).
(2) COUZENS, D. C. F. and TREVENA, D. H., *J. Phys.* D., 1974, **7,** 2277.

Part 2. Compressing a liquid

8.7. *Introduction*

Canton first demonstrated in 1762 that liquids were compressible. Since that time the physics of high pressures has developed into a very important subject. The most outstanding name in this field is that of Bridgman whose work will be described in the next section.

The only modulus that can be defined for a liquid is its bulk modulus, K, since as we saw in section 1.4, a liquid cannot support a shear stress. If we take a volume V of liquid and subject it to a hydrostatic pressure Δp the volume will change to $V - \Delta V$ and the volume strain will be $-\Delta V/V$. The bulk modulus is then

$$K = \frac{\Delta p}{-\Delta V/V}$$

$$= \frac{-V\Delta p}{\Delta V}.$$

The value of the bulk modulus depends on the rate at which the pressure changes are performed. If the pressure is applied slowly the liquid will remain at a constant temperature if it is contained in a vessel with conducting walls; in such a case we have the *isothermal* bulk modulus, K_{T}. If, on the other hand, the pressure changes occur so rapidly that practically no heat exchange with the containing vessel is possible in the time available we use the *adiabatic* bulk modulus K_{S}.

The *isothermal* compressibility β_T is defined as the reciprocal of K_T and is written as

$$\beta_T = -\frac{1}{V}\left(\frac{\partial V}{\partial p}\right)_T$$

where the suffix T emphasizes the fact that we are dealing wit isothermal conditions. It is this isothermal compressibility which i usually involved in measuring (p,V,T) data for liquids. The com pressibility of a liquid is usually small and numerically comparabl with that of a solid. For water at 15°C and for pressures in th range 1–25 atm the relevant data are:

$$K_T = 2{\cdot}05 \times 10^9 \text{ N m}^{-2} \text{ (Pa)}$$

$$\beta_T = 4{\cdot}9 \times 10^{-10} \text{ m}^2 \text{ N}^{-1} \text{ (Pa}^{-1}\text{)}.$$

8.8. *The experiments of Bridgman and others*

In measuring the bulk modulus of a liquid there are two particul difficulties involved. Firstly, the change of volume of the liqui when compressed is very small. Secondly, the expansion of th containing vessel is comparable with the change of volume of the liquid. Even if the same pressure were applied inside and outside the vessel the internal volume of the vessel would not remain constant but would decrease by the same amount as if a piece of the solid that exactly fitted the inside of the vessel were subjected to the same pressure on its outside. The expected volume change of a vessel can be estimated theoretically for vessels of simple geometrical shapes such as spheres or cylinders with flat ends. Metal vessels are best used because they are elastically isotropic.

In Bridgman's work the test liquid was placed in a strong alloy-steel cylinder and the apparent change in volume of the liquid relative to the cylinder measured by the advance of the piston producing the pressure. Bridgman ensured that the liquid did not leak upwards by using a special packing arrangement (see fig. 8.9). The piston A as it moves down transmits the pressure via the steel ring B to the piston C through soft rubber packing kept in place by copper washers D. Since the total upward force exerted by the test liquid on the lower face of C must equal that exerted by the packing the *pressure* in the packing is automatically greater than that in the liquid which therefore cannot leak. The pressure was determined by measuring the change

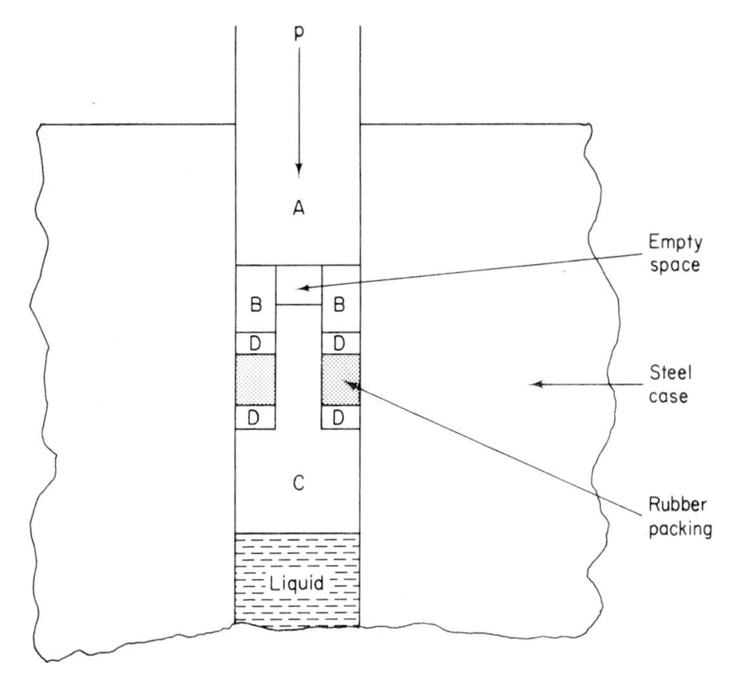

Fig. 8.9. Bridgman's apparatus.

in resistance of a coil of manganin wire contained in the cylinder. A correction for the expansion of the cylinder itself could be made by further experiments in which part of the liquid was replaced by a steel plug. With this apparatus Bridgman obtained (p, V) isotherms over a wide temperature range up to pressures of the order of 10 000 atm.

As the pressure is increased Bridgman found that the compressibility of liquids decreases quite considerably. For example the compressibility at a pressure of 12 000 atm is only about 1/20 of its value at fairly low pressures. This low compressibility at high pressures can be explained in terms of the intermolecular forces; as the molecules get ' squashed ' closer together the strong repulsions between them are brought into play and it is also thought that there is some shrinkage of the molecules themselves at very high pressures. The compressibility of most liquids increases as the temperature rises.

The effects of high pressure on the properties of liquids are many and varied. Such pressures can alter the transport properties and can produce chemical and biological changes.

Some important work on the compressibility of liquids has been recently carried out by Hayward.

References

(1) Bridgman, P. W., *The Physics of High Pressure* (G. Bell & Sons, London, 1949).

(2) *High Pressure Physics and Chemistry*, Volume 1, Edited by R. S. Bradley (Academic Press, London and New York, 1963).

(3) How to Measure the Isothermal Compressibility of Liquids Accurately, by A. T. J. Hayward, 1971, *J. Phys.* D, **4,** 938.

CHAPTER 9

viscosity

9.1. *Transport processes*

As the name implies, a transport process in a medium occurs when something is transferred from one part of the medium to another. The most familiar transport processes are viscosity and thermal conduction and they involve the transport of momentum and energy, respectively, through a medium. For gases it is possible, on the basis of the kinetic theory, to derive expressions for both the viscosity and thermal conductivity in terms of molecular properties and the reader is referred to text-books on properties of matter for the details. For a liquid, however, a corresponding derivation is not such an easy matter. But it is clear that, in either a gas or liquid, transport processes are *non-equilibrium* processes; and remember that, up to the present, we have been considering only liquids in thermodynamic *equilibrium.* In equilibrium a given property is assumed to be the same throughout the whole volume of the liquid and also not to change with time except for small fluctuations. Consider, however, what happens when a property—say temperature—is *not* uniform throughout a liquid at a given time. When this occurs heat energy is transported from the hotter to the colder regions and the resulting thermal conduction is a transport process in which energy flows in a direction opposite to that of the temperature gradient. Again consider viscosity, which comes into play when adjacent parts of the liquid move with different velocities, that is, a velocity gradient is present. The viscous forces involved act so as to cause the slower moving regions to move more rapidly and the faster moving ones more slowly. This tendency to ‘ even out ’ the velocities involves a transport of momentum in a direction perpendicular to the fluid velocity (see section 9.3). There is a third well-known transport process, namely, ordinary diffusion which is the transport of mass from one region to another because of a density gradient. In this chapter we shall be concerned only with viscosity.

9.2. *Viscous flow at constant rate of shear*

When an incompressible liquid is subjected to a constant shear it offers a resistance known as a viscous force or viscosity. This viscosity may be thought of as an internal friction between various layers of the liquid.

To illustrate these ideas let us consider the state of affairs shown in fig. 9.1 (a) where we have a region of liquid confined between two flat parallel plates each of area A at a vertical distance a apart. The plane of each plate is perpendicular to the plane of the paper which is taken to be the xy plane. Suppose that the lower plate is fixed and the top plate is dragged to the right in the x-direction with a constant velocity v by a force F. For many liquids it is found that the value of the force needed to maintain this constant velocity v is proportional to v. Such liquids are called *Newtonian* liquids. Most ordinary liquids with comparatively small molecules are Newtonian liquids, water being an example. For non-Newtonian liquids, usually possessing very large long molecules, this force is not proportional to the velocity v; such liquids are discussed in Chapter 10.

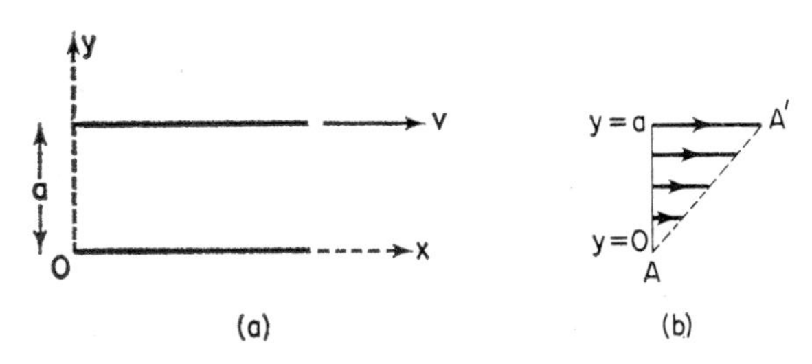

Fig. 9.1. (a) Viscous flow of a liquid between a lower fixed plate and an upper plate dragged with constant velocity v parallel to it. (b) The velocity profile AA′ corresponding to the flow in (a).

It is also assumed that there is no slip at either the top or bottom plates. This is because the liquid layer adjacent to each plate is 'adsorbed' to the surface owing to intermolecular attractions. Thus the layer of liquid adjacent to the bottom plate is stationary. It is further found that v_x, the x-component of the velocity of the liquid, varies linearly with the vertical distance y above the bottom plate; at $y=0$, $v_x=0$ and at $y=a$, $v_x=v$. If we draw horizontal vectors whose lengths are proportional to the values of v_x at various heights their ends will lie on the straight line AA′ in fig. 9.1 (b); this straight line is called the *velocity profile*. The streamlines are horizontal and this type of steady flow is called *laminar* or *streamline* flow. The velocity gradient in the y-direction is v/a and the value of v_x at any height y is $v_x=vy/a$.

The upper plate exerts a shearing stress $\tau=F/A$ in the x-direction on the top layer of the liquid and this stress is related to the velocity gradient by the equation

$$\tau=\eta\frac{v}{a}. \qquad (9.1)$$

This equation defines the coefficient of shear viscosity, η, also called the ' dynamic viscosity ' or simply ' viscosity '.

If we consider two layers of liquid at heights y and $dy+y$ above the bottom plate with velocities v_x and v_x+dv_x then the upper layer will be slipping over the lower one; just how easy it is for this slipping to occur depends on the ' internal friction ' between the layers, that is, on the viscosity. This upper layer will also exert a shearing stress $\tau=F/A$ on the lower layer and the velocity gradient is $dv_x/dy=v/a$. In fact the shearing stress on *any* layer of liquid parallel to the x-axis is $\tau=\eta v/a$ which we can also write as

$$\tau=\eta\frac{dv_x}{dy}.$$

In the type of flow we are considering there are two perpendicular directions involved; these are the direction of the flow velocity v_x (the x-direction) and the direction along which v_x varies (the y-direction). To emphasize this we can rewrite the last equation as

$$\tau_{xy}=\eta\frac{dv_x}{dy} \tag{9.2}$$

where the first suffix in τ_{xy} indicates the direction of the shear stress and the second the normal to the plane in which this stress acts.

The viscous forces brought into play are such that they tend to slow down the faster moving layers and speed up the slower moving layers. This will be discussed more fully in the next section.

η has the dimensions $[ML^{-1}T^{-1}]$ and is expressed in newton second per metre squared (N s m^{-2}). The quantity $\nu=\eta/\rho$ is also often used where ρ is the density of the liquid; ν is called the *kinematic viscosity* and its unit is the metre squared per second (m^2 s^{-1}). For water at 20°C the relevant values are

$$\eta=1{\cdot}002\times10^{-3}\ \text{N s m}^{-2},$$
$$\nu=1{\cdot}004\times10^{-6}\ \text{m}^2\ \text{s}^{-1}.$$

9.3. *Transfer of momentum in liquids*

We first adopt a ' macroscopic ' approach and consider the laminar flow of a simple Newtonian liquid whose molecules interact by means of central forces only (that is forces which depend only on the separation of the molecules, see section 4.1).

We return to consider fig. 9.1 (*a*). As the top plate moves to the right it imparts momentum to the liquid in contact with it. Since force is the rate of change of momentum, the momentum given to this top layer of liquid in time δt is $F\delta t$, where F is the force acting on the upper plate. The lower fixed plate will experience an equal and opposite force F because of the dragging force due to the liquid; thus the liquid in contact with this lower plate will be deprived of a momentum of $F\delta t$ in this time δt. Thus the resultant effect is as though a momentum of $F\delta t$ is transferred a distance a upwards through the liquid in the y-direction between the plates in a time δt. This transfer of momentum is (magnitude of momentum transferred) × (velocity of transfer), that is,

$$F\delta t \times \frac{a}{\delta t}$$

$$= Fa$$

$$= \tau_{xy} Aa$$

$$= \tau_{xy} V'$$

where V' is the volume of liquid between the plates.

We can also look at the problem from a more 'microscopic' point of view. Each molecule of the liquid is enclosed in a cell formed by its nearest neighbours and only migrates *very* occasionally to a neighbouring cell. In its cell it is under the influence of its intermolecular interactions with its neighbours. The effect of these interactions is to give rise to a 'drag' which causes the faster molecules to be slowed down while the slower ones are speeded up. To illustrate this mechanism consider two molecules i and j at a distance r apart as in fig. 9.2. The force between these molecules along the line joining them is $F(r)$ and the x-component of this force is $F_x = F(r) \cos \theta$. If we concentrate on molecule j then F_x is the rate of change of x-momentum of this molecule j. Thus in a time δt molecule j will suffer a change of $F_x \delta t$ in its x-momentum due to the interaction of molecule i. Similarly, by Newton's third law, this molecule i will suffer an equal but opposite change in x-momentum in time δt due to molecule j. Thus we can regard this mechanism as being equivalent to a transfer of x-momentum of $F_x \delta t$ through space from molecule i to molecule j in time δt, that is through a vertical

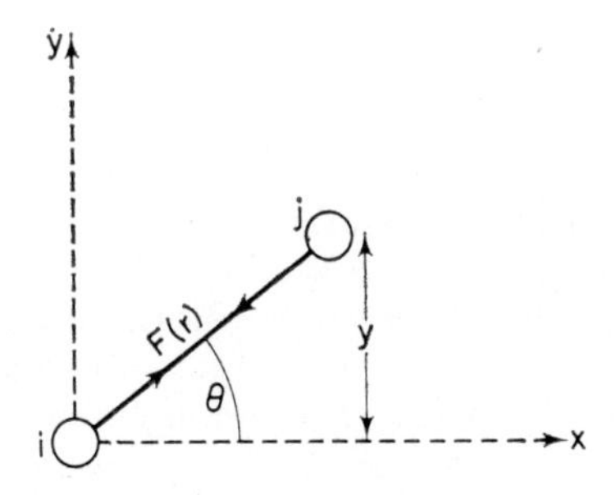

Fig. 9.2. Momentum transfer arising from the interaction of two molecules

distance y in time δt. Since molecule j is at a greater height above the fixed plate in fig. 9.1 (a) its velocity is greater than that of molecule i and the direction of $F(r)$ between the two molecules ensures that the x-momentum of molecule j is reduced whilst that of molecule i is increased. The rate of transfer of x-momentum in the y-direction is

$$\begin{aligned} F_x \delta t \times \frac{y}{\delta t} & \\ &= F_x y \\ &= F(r) \cos\theta \times y. \qquad (9.3) \end{aligned}$$

To calculate the total flow of x-momentum we must sum the expression (9.3) over all the ij pairs of molecules in the volume V' of liquid between the plates in fig. 9.1 (a). For further details the reader is referred to Chapter 9 of *The Liquid State* by Pryde.

Finally it is of interest to compare the molecular mechanism responsible for momentum transfer in a liquid with that in a gas. In a gas a molecule spends most of its time in empty space between collisions and hence the transport of momentum is accomplished by each molecule actually carrying its momentum with it as it thus moves. On the other hand, in a liquid, we have seen that the momentum transfer is due to the ' drag ' produced by the intermolecular interactions.

9.4. *Flow through tubes*

In practice we are usually concerned with the flow of a liquid through a pipe and in this section we shall consider the types of flow that can occur under a pressure gradient in a cylindrical tube.

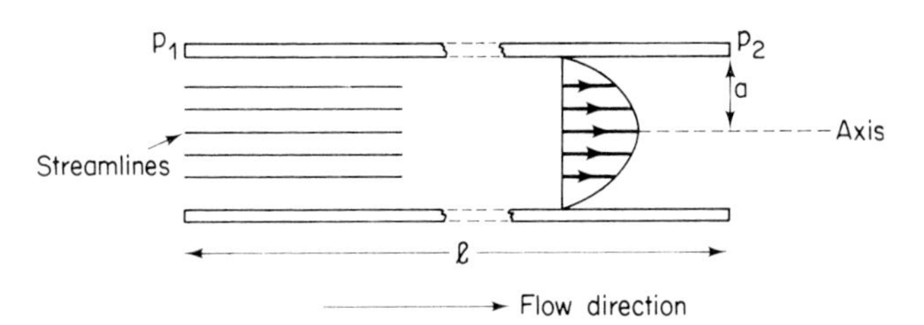

Fig. 9.3. Parabolic velocity profile when the flow is streamline throughout the liquid.

Consider a horizontal cylindrical tube of internal radius a and length l. Liquid enters one end of the tube at a pressure p_1 and leaves it at the other at a pressure p_2 (fig. 9.3).

We first consider streamline flow. It is assumed that the layer of liquid in contact with the cylindrical tube wall is at rest and that there is also no radial flow of the liquid; the streamlines are therefore all parallel to the axis of the tube. It can then be shown that the velocity of flow at a radial distance r from the axis of the tube is given by

$$v = \frac{(p_1 - p_2)(a^2 - r^2)}{4\eta l}. \tag{9.4}$$

This shows that v varies from zero at the tube wall ($r = a$) to a maximum at the axis ($r = 0$) and the velocity profile across the tube is a parabola whose axis coincides with that of the tube (fig. 9.3).

It can further be shown that the volume flow Q per second of liquid through the tube is given by

$$Q = \frac{(p_1 - p_2)\pi a^4}{8\eta l} \tag{9.5}$$

a result known as Poiseuille's equation.

The maximum velocity of flow is given by substituting $r = 0$ in equation (9.4), whence

$$v_{max} = \frac{(p_1 - p_2)a^2}{4\eta l}.$$

Thus equation (9.5) can be written as

$$Q = \pi a^2 \left(\frac{v_{max}}{2}\right).$$

If we divide the volume of liquid Q flowing per second across a cross-section of the tube by πa^2 this defines the *mean* velocity of flow $\bar{v}$. Thus

$$\bar{v} = \frac{Q}{\pi a^2} = \frac{v_{max}}{2}.$$

So in streamline flow the mean velocity of flow is half the maximum flow velocity.

In the above treatment two important matters have been overlooked but in a more rigorous approach they would have to be included. The first is the 'kinetic energy correction' which arises from the fact that the pressure difference $(p_1 - p_2)$ imparts kinetic energy to the moving liquid in addition to overcoming the viscous forces. The second effect which has been neglected is that the flow of the liquid is not strictly streamline along the *whole* length of the tube. For a short distance along the inlet end of the tube there is some radial flow which disappears at the point where the true streamline flow sets in. These two effects can be allowed for in a more rigorous treatment.

If the flow velocity in the tube is fairly small then we have the steady streamline flow just considered. As the flow velocity is increased this streamline flow will persist until, at a certain critical value v_c of the mean flow velocity, a complete change in the flow pattern occurs and we say that *turbulence* sets in. The situation is best described by reference to fig. 9.4 where a portion of the tube is shown.

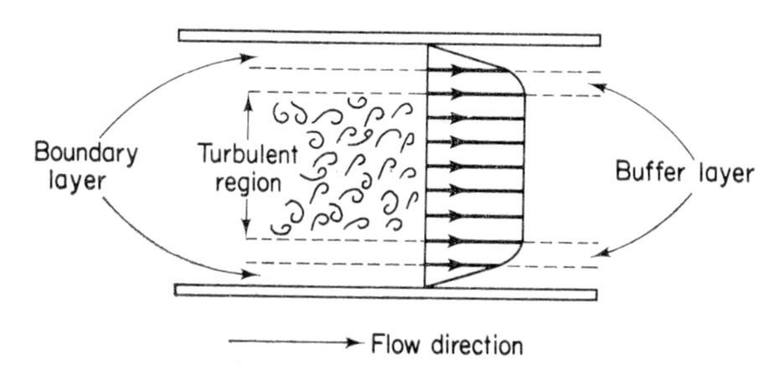

Fig. 9.4. Velocity profile in the case where the main stream is turbulent and with streamline flow in the boundary layer.

In this turbulent flow the layer of liquid in contact with the tube wall will still be stationary. Just inside this wall there is a narrow ' boundary layer ' in which the liquid flow is still streamline; in this boundary layer the flow velocity increases uniformly as we go in from the wall of the tube. Inside the boundary layer there is a narrow ' buffer ' or transition layer and finally we reach the main inner region where the flow is turbulent. The flow velocity has a constant mean value in this inner turbulent region as shown by the velocity profile in fig. 9.4; this is in marked contrast with the parabolic distribution characteristic of Poiseuille flow (fig. 9.3). In the turbulent region there is a thorough mixing of the liquid due to a large number of eddy currents or *vortices*.

Four parameters determine whether the flow in a cylindrical tube is streamline or turbulent. These are the density ρ of the liquid, the tube diameter d, the mean flow velocity v' and the viscosity η. They can be combined to give a dimensionless quantity Re known as *Reynolds' number*, defined by

$$Re = \rho v' d/\eta. \tag{9.6}$$

Re is really a measure of the relative strengths of the inertial and viscous forces in the liquid.

The value of Re corresponding to the critical velocity $v' = v_c$ is about 2000. For $Re < 2000$ or so the flow is streamline and when Re exceeds this value turbulence sets in. If we assume that turbulence in water at 20°C occurs at $Re = 2000$ then, from equation (9.6), the value of the mean flow velocity turns out to be 1 m s^{-1} for a tube of diameter 2 mm.

One final word about the energy aspect of the two types of flow. For streamline flow, the work done by the shearing forces 'overcomes internal friction', increases *molecular* kinetic energy and so generates heat. In turbulent flow, most of the work done organizes a large number of small eddies, with kinetic energy on the macroscopic scale.

9.5. *Viscosity and rigidity: elongational viscosity*

We now discuss the analogy between the rigidity of an elastic solid and the viscosity of a flowing liquid.

For an elastic solid the shear modulus G is given by

$$G = \frac{\text{shear stress}}{\text{shear strain}}. \tag{9.7}$$

For a simple Newtonian liquid the viscosity η is given by equation (9.1), that is,

$$\eta = \frac{\text{shear stress}}{\text{velocity gradient}}. \tag{9.8}$$

Consider a cube of an elastic solid (shown end-on) which has been sheared as shown in fig. 9.5 with the upper plane B displaced through a distance v with respect to the lower plane A in one second. The shear produced in this second is v/y.

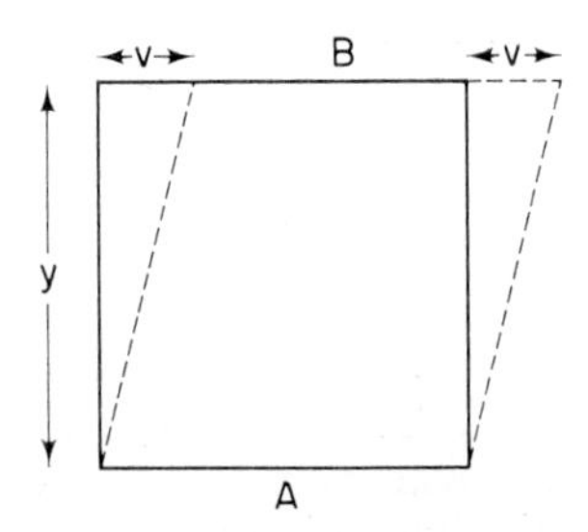

Fig. 9.5. Cube of elastic solid, viewed end-on, under shear strain.

Next suppose that A and B are parallel planes in a Newtonian liquid in an arrangement like that in fig. 9.1 (*a*) with plane A at rest and plane B moving a distance v in one second, that is, v is the velocity of B with respect to A. There will be no shear strain in the liquid since the liquid flows so as to destroy the shear strain that would have been produced in the absence of flow. So in the liquid a shea strain of v/y is destroyed per second, that is, v/y is the rate of change (or disappearance) of shear strain. v/y is also the velocity gradient. Thus:

$$\text{velocity gradient} = \text{rate of change of shear strain.} \tag{9.9}$$

Thus equation (9.8) can also be written as

$$\eta = \frac{\text{shear stress}}{\text{rate of change of shear strain}\dagger}. \tag{9.10}$$

From equations (9.7) and (9.10) we see that dimensionally

$$[G] = \left[\frac{\eta}{\text{time}}\right]. \tag{9.11}$$

For an elastic solid wire the Young modulus E is given by

$$E = \frac{\text{tensile stress}}{\text{longitudinal strain}}. \tag{9.12}$$

When a 'rod' of liquid (like pitch) is stretched by a tractile (i.e. tensile) force its length increases, that is, the rod flows along its length. This enables us to define the *elongational viscosity* λ as

$$\lambda = \frac{\text{tensile stress}}{\text{rate of change of longitudinal strain}}. \tag{9.13}$$

Dimensionally we have, from (9.12) and (9.13),

$$[E] = \left[\frac{\lambda}{\text{time}}\right]. \tag{9.14}$$

So for a solid we have G and E; for a liquid we have η and λ.

For a solid G and E are related by

$$E = 2(1+\sigma)G \tag{9.15}$$

where σ, Poisson's ratio, is defined for a stretched solid wire as

$$\sigma = \frac{\text{diameter change/original diameter}}{\text{longitudinal strain}}$$

$$= \frac{\mathrm{d}r/r}{\mathrm{d}l/l}.$$

For elastic solids $\sigma < 1/2$ and this means that there is an increase in volume when such a solid rod is stretched.

† The denominator is also often referred to as the 'rate of shear'.

In general, when a cylinder of any material has its length changed from l to $(l+\mathrm{d}l)$ its radius changes from r to $(r-\mathrm{d}r)$ and the volume change $\mathrm{d}V$ is

$$\begin{aligned}\mathrm{d}V &= \pi(r-\mathrm{d}r)^2(l+\mathrm{d}l)-\pi r^2 l \\ &= \pi r^2\mathrm{d}l - 2\pi r l\mathrm{d}r\end{aligned}$$

neglecting second order terms.

If the cylinder is made of incompressible material then $\mathrm{d}V=0$, that is,

$$2\pi r l\mathrm{d}r = \pi r^2\mathrm{d}l$$

or

$$\frac{\mathrm{d}r/r}{\mathrm{d}l/l} = \sigma = \tfrac{1}{2}.$$

So a material for which $\sigma=\frac{1}{2}$ suffers no change in volume during an extension or compression. For such a material, if it is a solid,

$$E=3G \tag{9.16}$$

from (9.15).

Since $[E]=[\lambda/\text{time}]$ and $[G]=[\eta/\text{time}]$ then it follows that

$$\lambda=3\eta \tag{9.17}$$

also for an incompressible liquid for which $\sigma=\frac{1}{2}$. A Newtonian liquid is such an incompressible liquid.

References

(1) Cottrell, A. H., *The Mechanical Properties of Matter*, Chapter 12 (John Wiley & Sons, Inc., London, 1964).

(2) Sprackling, M. T., *The Mechanical Properties of Matter*, Chapter 5 (The English Universities Press Ltd., London, 1970).

(3) Stanley, R. C., *Mechanical Properties of Solids and Fluids*, Chapter 4 (Butterworth & Co. Ltd., London, 1972).

(4) Noakes, G. R., *New Intermediate Physics*, Chapter 9 (Macmillan & Co. Ltd., London, 1957).

CHAPTER 10

rheology

10.1. *Introduction*

' Rheology ' is the study of the deformation and flow of matter and this chapter will be devoted to a discussion of the main types of rheological behaviour.

In section 9.2 we described the flow of a liquid for which the coefficient of viscosity η is a constant, independent of the shear rate, for a particular liquid at a given temperature and pressure. This viscosity is given by

$$\eta = \frac{\text{shear stress} \quad \tau}{\text{rate of shear} \quad S}$$

(see section 9.5).

This equation is Newton's law of constant viscosity and liquids which are described by it are known as *Newtonian* liquids. For these liquids the graph of shear stress against rate of shear is a straight line through the origin, so long as the flow is laminar (see curve 1 of fig. 10.1). Such a curve of shear stress against rate of shear is known as the *consistency curve* for the liquid.

Examples of Newtonian liquids are water, most pure single-phase liquids and solutions of substances of low molecular weights. The viscous dissipation of energy in such liquids is due to collisions between *fairly small* molecules. Typical values of viscosity at 20°C are: water 1·002, acetone 0·324, mercury 1·552 mN s m^{-2}; the viscosity of air at 20°C and normal pressure is, for comparison, 18·2 μN s m^{-2}.

Several liquids do not have a constant coefficient of viscosity and their viscosity varies with the rate of shear. Such liquids are called *non-Newtonian* and may be divided into two groups known as time-independent and time-dependent liquids. These two groups of liquids will be described in the next two sections.

10.2. *Time-independent non-Newtonian liquids*

These non-Newtonian liquids are ones whose viscosity is not dependent on their duration of flow. (Newtonian liquids are also time-independent.) The consistency curves for the various types of liquid in this category are shown in fig. 10.1.

As already mentioned, curve 1 of fig. 10.1 is the straight line corresponding to a Newtonian liquid and its slope gives the value of

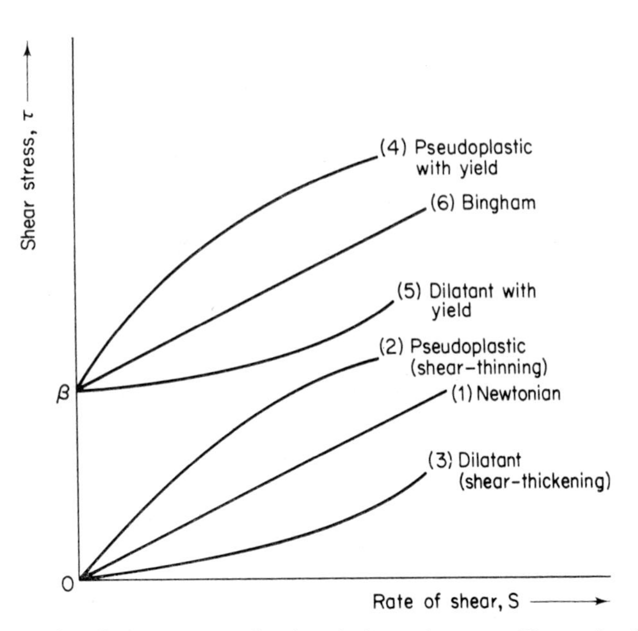

Fig. 10.1. Consistency curves for time-independent non-Newtonian liquids.

the constant viscosity η of this liquid. Curve 2 is the consistency curve for a liquid known as a *pseudoplastic* or *shear-thinning* liquid. This curve has no linear portion, that is, the value of τ/S varies along it so that there is no coefficient of viscosity in the sense which applies to a Newtonian liquid. Nevertheless if we take any point of coordinates (τ, S) on this curve the ratio τ/S gives the value of the *apparent viscosity*, η_a, at this point. When a value of the apparent

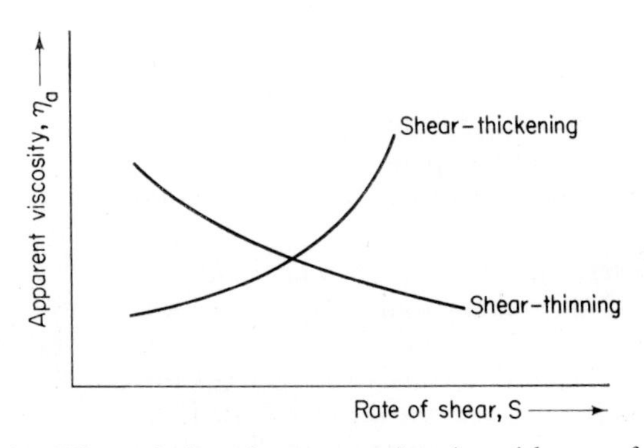

Fig. 10.2. The variation of apparent viscosity with rate of shear for shear-thinning and shear-thickening liquids.

viscosity is quoted it is meaningless unless the rate of shear is also stated, since η_a depends on the shear rate and varies as we move along curve 2. These liquids thin upon shearing, that is, their apparent viscosity decreases with increasing shear rate (see fig. 10.2).

Examples of shear-thinning liquids are dilute solutions of high polymers, certain polymer melts such as rubbers and cellulose and some suspensions such as paints and detergent slurries.

A possible explanation of the behaviour of shear-thinning liquids is as follows. Take, for example, a dilute high polymer solution. As the rate of shear increases it is thought that the long polymer molecules become increasingly aligned along the streamlines in much the same way that molecular dipoles become aligned by an increasingly applied magnetic field. As this degree of alignment increases the apparent viscosity becomes progressively less and when the molecular major axes coincide with the streamlines the apparent viscosity attains a steady value thereafter.

Curve 3 is the opposite of curve 2 and shows the behaviour of a *dilatant* or *shear-thickening* liquid. For these liquids the apparent viscosity increases with increasing shear rate as shown in fig. 10.2. These liquids thicken upon shearing and are far less common than the shear-thinning types. A good example of a shear-thickening liquid is one described by Collyer (1973 b) in which 66·7 g of wheat starch is mixed with 100 cm^3 of water. When this mixture is stirred slowly with a rod, the viscous force on the rod is small. If however the rod is moved suddenly, so as to involve a high shear rate, the viscous force arrests the motion of the rod. Again, such a rod will fall easily under gravity through the liquid but if it is suddenly plunged into the surface of the liquid its motion is very soon arrested. At present there is no satisfactory explanation for the mechanism responsible for shear-thickening.

A plot of log (*shear stress*) against log (*shear rate*) is usually linear for liquids described by curves 2 and 3, that is, the relation

$$\tau = \alpha S^n$$

holds, where α and n are constants for a particular liquid. For a shear-thinning liquid $n < 1$ and for a shear-thickening liquid $n > 1$. (For a Newtonian liquid $n = 1$.) The apparent viscosity is given by

$$\eta_a = \frac{\tau}{S}$$

$$= \alpha S^{n-1}.$$

Since, for a shear-thinning liquid, $n < 1$ then η_a decreases as S increases in agreement with fig. 10.2. Similarly since $n > 1$ for a shear-thickening liquid, η_a increases as S increases.

Some pseudoplastic materials deform initially like solids until a certain value β of the shear stress is exceeded; this value β is known as the *yield stress*. No flow will occur for shear stresses less than this value but for higher shear stresses these materials behave as normal pseudoplastic liquids. Such a material is called a *pseudoplastic with yield value* and is represented by the curve 4 in fig. 10.1. The flow stops when the shear stress falls below the yield value β. A pseudoplastic with yield value is, in behaviour, a strange mixture of a solid and a liquid: for shear stresses below β it behaves as a solid and for shear stresses above this yield value it behaves as a pseudoplastic liquid. In the same way there are materials known as *dilatant with yield value* and these are represented by the curve 5 of fig. 10.1; such a material also behaves as a solid for stresses below the yield value.

One special case of a material with a yield value is known as a *Bingham plastic* whose behaviour is shown by the straight-line curve 6 of fig. 10.1. This material deforms elastically until the yield stress β is reached and once this stress is exceeded it flows as a Newtonian liquid with τ linearly related to S. So a Bingham material may be described as Newtonian with a yield value.

For any point (τ, S) on the straight line 6 of slope μ we can write

$$\mu = \frac{\tau - \beta}{S}$$

where μ is the Bingham viscosity for a yield value of β. The corresponding apparent viscosity of the Bingham plastic is

$$\eta_a = \frac{\tau}{S} = \mu + \frac{\beta}{S}. \qquad (10.1)$$

The Bingham plastic is a very useful 'ideal' theoretical model and there are several materials whose actual behaviour approximates closely to it. Fresh cement is one such material and typical values of μ and β are 2400 mN s m^{-2} and 48 N m^{-2} respectively.

Consider a material with a yield value moving along a capillary. The maximum shear stress on the material occurs at the walls and this stress decreases as we go in along a radius and becomes zero along the axis of the tube. As the shear stress is gradually increased the following sequence of events occurs. At a certain value, τ_1, of shear stress the static friction at the wall of the capillary is overcome and the material moves as a solid plug along the capillary. As the shear stress is further increased then, at first, a narrow annular region of streamline flow will occur near the wall. Over this region between $r = r_c$ and $r = r_0$ the stress will have exceeded the yield value of the material; this is the shaded part in fig. 10.3. For values of $r < r_c$ the stress will be less than the yield value and the material moves as a solid plug of radius r_c. As the shear stress is further increased the yield value will be exceeded at points nearer the axis of the tube and the size of the central solid plug gets less and less.

A good example of Bingham plastic flow is the flow of toothpaste out of its tube and it will be readily appreciated that the yield value β is essential to the action of the toothpaste. Other examples are drilling muds and fresh cement. The usual explanation of Bingham plastic behaviour, and of close approximations to it, is as follows. These materials have a three-dimensional structure which is capable of resisting shear stresses as high as β; once the shear stress exceeds β this structure breaks down and flow starts to occur. Reducing the shear stress subsequently to values below the yield value β causes the structure to reform.

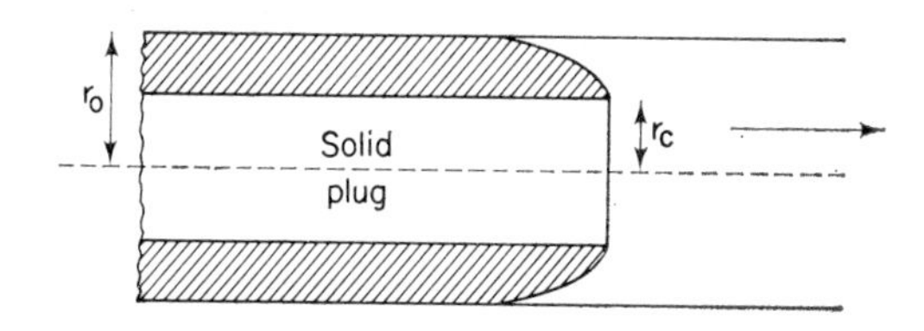

Fig. 10.3. Illustrating the central solid plug in the flow of a material with yield value along a capillary.

10.3 *Time-dependent non-Newtonian liquids*

Time-dependent liquids are ones whose viscosity depends on the luration of flow. For these liquids there is no *unique* relation etween shear stress τ and rate of shear S and so consistency curves annot be drawn to represent the behaviour of such substances.

When some liquids of this kind are made to flow under a constant ate of shear the shear stress and the apparent viscosity gradually ecrease with time. Such liquids become 'runny' when stirred or haken and form a *sol*. On then being left alone they reset, forming *gel*. The name given to such a reversible isothermal gel–sol–gel equence is *thixotropy*. Salad cream and ketchup are examples of his kind of substance. The undisturbed liquid in the bottle is a gel ut after shaking the bottle and thus shearing the contents the iscosity of the liquid decreases sufficiently for one to be able to pour out. Thixotropic paints are also well known; these paints, after eing stirred to a sol, can easily be applied to a surface where they oon reform into a gel which will not 'run'. Thixotropy is also of nportance in certain drilling muds.

Some thixotropic materials behave like a Bingham plastic in the sense that they have a yield value below which no flow occurs; they are called *thixotropic-plastic* materials.

Thixotropic liquids revert to their original structure on being left to stand. For some liquids however this 'rebuilding' of structure occurs more rapidly when the liquids are stirred gently or their containers rolled slowly in the hands. Following Collyer (1974) we shall call these *rheopectic* liquids. They are essentially thixotropic liquids which also rebuild their original structures more quickly if helped along by a little gentle shearing. Small shearing motions in rheopectic liquids help to rebuild the structure once it has been destroyed while larger shears destroy the structure. There is a critical value of the shear above which the structure tends to be destroyed rather than rebuilt. All thixotropic liquids, by definition, regain their initial structures after shearing.

Freundlich and Juliusberger in 1935 carried out some interesting experiments with a rheopectic liquid. This liquid consisted of a 42 per cent gypsum paste in water. It was found that if this liquid were shaken into a sol it reset in about 40 min if allowed to stand undisturbed. If instead its container was rolled gently in the hands the resetting process occurred in the much shorter time of 20 s.

Negative thixotropic liquids are the reverse in their behaviour to thixotropic liquids and are much less common. Their apparent viscosity increases with time under a constant rate of shear and they exhibit a sol–gel–sol sequence which is reversible and isothermal. (The term rheopectic has been used for a negative thixotropic liquid but this only causes confusion; we shall use this term to describe the special kind of thixotropic liquids mentioned earlier.) A good example of negative thixotropic behaviour has been given by Cheng (see reference at the end of this chapter).

Since there are several different types of thixotropic materials like sols, suspensions, polymer solutions and melts there is no one model which will satisfactorily account for the behaviour of them all. So we shall only describe a *general* kind of model which will explain the main features of thixotropic behaviour. The large molecules of which thixotropic materials are generally composed are usually packed fairly loosely because of their unsymmetric shapes—thin discs, long needles and polymer chains. A proposed model for a thixotropic-plastic material (that is, one with a yield value) must explain both the existence of this yield value and also the time-dependence of the subsequent flow. The first requirement is met by assuming that there is some form of thixotropic bonding between the molecules which keeps the structure rigid up to the yield value. Once the flow starts the continued decrease in apparent viscosity is thought to be due to the progressive alignment of the major axes of the molecules along the streamlines. In these thixotropic liquids this alignment requires a finite and measurable time; this is in contrast with the case of the shear-thinning liquids we discussed in section 10.2 where such alignment occurs in an unmeasurably small time.

In negative thixotropic liquids it is thought that *more* intermolecular bonds are formed as a result of stirring motions as compared with the number in the liquid at rest. This increase in the number of bonds on shearing explains how a gel can be produced.

10.4. *Applications of non-Newtonian liquids*

In the food industry there are many cases where the properties of non-Newtonian liquids are of relevance and we quote a few examples. The yield value of margarine is important since the material is required to be fairly firm before use but must flow when spread on bread. The yield values and flow properties of ketchup and salad cream are also of importance. Chocolate and fruit gums must be firm at ordinary room temperatures but must flow in the warmer temperature in the mouth.

Toothpaste and shaving creams have to stay in their tubes but must be able to flow out when these tubes are squeezed. It is necessary that thixotropic paints, in the sol state, should spread easily on a surface but they must also quickly revert to the gel state and not run over the surface. The yield value for ink in a ball-point pen is of importance since the ink must not run except when the pen is used in writing.

In oil drilling work the fluid used to lubricate the drill is called a drilling mud. In drilling a well this mud is pumped down inside the drill along its axis and it is then forced upward in the annular gap between the drill and the cylindrical wall of the well. The main function of the mud is to lift drilling debris from the drill head to the top of the well. The yield value is of great importance here: if the drill is stopped the drilling debris will tend to fall down the well unless the mud sets at once (as in a Bingham type mud) or at least fairly quickly as in a thixotropic mud of the right consistency. What is really required is that the debris remain in suspension and not settle down at the bottom of the well.

For further details of these applications of non-Newtonian liquids the reader is referred to the articles by Collyer listed at the end of the chapter.

10.5. *Viscoelastic liquids*

This is another class of liquids which possess elasticity as well as viscosity. The 'classic' example is perhaps the material known

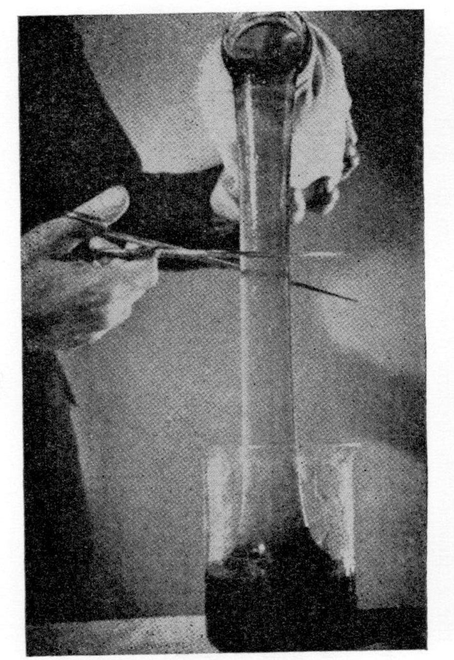

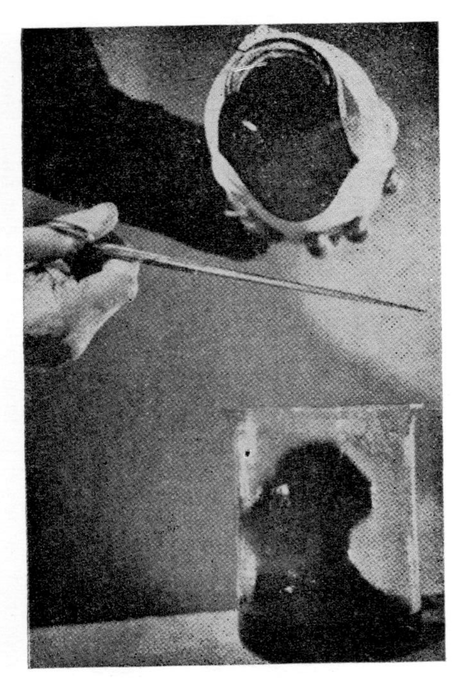

Fig. 10.4. Elastic recovery in a liquid. A solution of aluminium dilaurate in a hydrocarbon oil is poured from a bottle (*left*); the pouring is stopped, and the stream cut (*centre*); the upper stream retracts quickly towards the mouth of the bottle, and the lower stream decreases in length and increases in width (*right*). (Photos by R. A. Barker.)

as ‘ Silly Putty ’ (polydimethylsiloxane). A small sphere of this putty will flow and flatten under its own weight in a short period (a quarter of an hour or so) if left to itself; this is due to its viscous flow. On the other hand the same sphere will readily bounce with a coefficient of restitution of about 0·7 thus showing that it has marked elastic properties as well.

Many solutions and melts of high polymers, with very long molecules (macromolecules), are viscoelastic. As in the case of the liquids discussed in sections 10.2 to 10.4 the viscous behaviour will be related to the degree of alignment of the molecules with the streamlines and will depend on the bonds between the macromolecules. The elastic behaviour is related to the internal structure of the macromolecules; each such molecule has a number of ‘ rigid ’ links and each link can take up various orientations with respect to its neighbours thus enabling each molecule to assume a large number of shapes as the material is deformed elastically.

In the next section we shall discuss the best-known phenomena exhibited by viscoelastic liquids. In doing so we shall concentrate in particular on solutions of polyethylene oxide because these solutions can be easily prepared and exhibit most of these effects very satisfactorily.

10.6. *Phenomena exhibited by viscoelastic liquids*

(*a*) *Elastic recovery*

When a viscoelastic liquid is stirred any air bubbles in it will retrace their motions when the stirring is abruptly stopped. This recoil movement of the bubbles is due to the so-called *elastic recovery* of the liquid and the actual extent of this recoil is a measure of the elasticity of the liquid. Again, if the vessel containing this liquid is given a sudden twist the liquid inside will oscillate back and fore for some seconds.

Another example of elastic recovery is obtained when a viscoelastic liquid is poured from one beaker into another. If the pouring is stopped for a few seconds while the vertical liquid column is cut sharply with a scissors the upper part of the column quickly ‘ recovers ’ by recoiling suddenly into the upper beaker (see Figure 10.4).

(*b*) *Spinability and self-siphoning*

It is possible to draw a polymer solution into long, thin threads, a process called *spinability* whose existence is the basis of the manufacture of nylon threads. The greater the polymer concentration the more stable are the liquid threads.

A viscoelastic liquid, in the form of a continuous filament, can be made to flow over the lip of a beaker even though the surface of the

Fig. 10.5. Self-siphoning effect using a 0·8% solution of polyethylene oxide (after Collyer).

liquid is below the rim. Fig. 10.5 shows this self-siphoning effec
If a vessel of such a liquid is tipped, it is very difficult to stop the flov
since the siphoning effect empties the vessel even when it is returne
to an upright position.

(c) The Weissenberg effect

Weissenberg first discovered this phenomenon very simply b
rotating a vertical cylindrical rod about its axis in a viscoelastic liqui
He found that the liquid climbed up the rod for quite a distan
while the rod was rotating in this way. (By contrast, if a Newtoni

liquid is stirred in a beaker the liquid rises up the walls of the beaker, an effect well known to anyone who has stirred a cup of tea.) Collyer has shown this Weissenberg effect even better by floating a 0·6 per cent polyethylene oxide solution (the viscoelastic liquid) on the surface of dimethyl phthalate and covering the viscoelastic liquid with a layer of Tellus 15 oil. When a glass rod is rotated in this three-layered arrangement the viscoelastic liquid flows up into the oil and also downward into the dimethyl phthalate as shown in fig. 10.6.

Fig. 10.6. The Weissenberg effect. The middle liquid, which is viscoelastic, encroaches into both Newtonian liquids (after Collyer).

The explanation of the Weissenberg effect is as follows. As the rod rotates the streamlines in the viscoelastic liquid are horizontal concentric circles whose centre lies on the axis of rotation of the rod. Now it is known that in a viscoelastic liquid the stresses in the direction of the streamlines and in a direction normal to the streamlines (in this case this is the vertical direction) are not equal. The difference between these stresses, known as the normal-stress *difference*, causes the viscoelastic liquid in fig. 10.6 to climb up and down the rotating rod. These normal-stress differences, which are always present in the flow of a viscoelastic liquid, are not present in a Newtonian liquid.

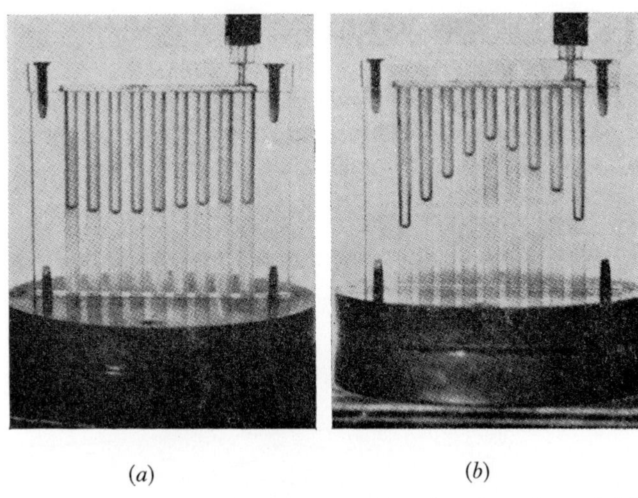

Fig. 10.7. Pressure distribution generated by the shear flow of (*a*) a Newtonian liquid and (*b*) a viscoelastic liquid between a fixed upper plat (with manometers) and a parallel rotating plate (not visible).

Another illustration of these normal-stress differences is obtaine if a viscoelastic liquid is sheared between two parallel circular plate which are in relative rotation. Arrangements of this kind are show in fig. 10.7. In both cases the upper plate is fixed and the lower plat (not shown) is rotating; the radial distribution of pressure on the uppe plate is measured by means of the vertical tube manometers as show For the first liquid, which is Newtonian, we see the effects due t centrifugal forces in the absence of any normal-stress difference In this liquid the pressure is greatest at the rim due to the action of th centrifugal forces alone and the manometer levels show this. Fo the second liquid, which is viscoelastic, the manometer readings a quite different and show how the normal-stress components vary alon a given horizontal radius, showing clearly that this component greatest at the centre.

(d) The die-swell effect

Consider a viscoelastic liquid flowing downwards through a tube of circular cross-section. Then (unlike the case with a Newtonian liquid), stresses are set up normal to the streamlines which are parallel to the axis of the tube. When the liquid emerges at the lower end of the tube these normal stresses are relieved by an increase in diameter of the emerging stream, sometimes to as much as three or four times the diameter of the tube (fig. 10.8). This increase of diameter is referred to as a *die-swell* and was first observed by Barus in 1893.

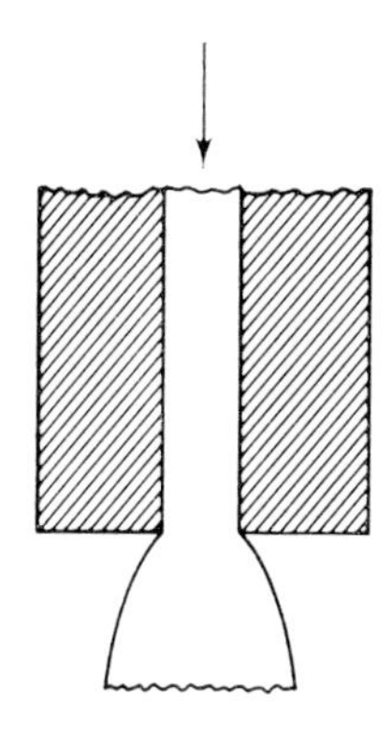

Fig. 10.8. The die-swell effect.

Die-swell is a common feature in the extrusion of man-made fibres and plastics. Increases in diameter occur in rayon spinning, so that t is essential to know beforehand the amount of die-swell likely to occur when a fibre of a specified diameter is required.

e) Drag reduction and jet stabilization

In the streamlined flow of a liquid through a pipe the viscous drag balances the force which causes the liquid to flow through the pipe. When turbulence sets in, as the flow rate increases, the viscous drag ncreases more rapidly than the rate of flow. By adding a small uantity of polymer to the liquid the viscous drag is reduced and it is hen possible to obtain a greater flow rate for a given pressure ifference across the ends of the tube. This is called *drag reduction.* The addition of a polymer can also produce *jet stabilization* in jets of vater. Normally at a certain distance from the nozzle the water jet tream starts to break up but addition of the polymer delays this rocess. These two effects are simultaneously put to good use by remen; they enable jets of greater height and of a more stable nature be produced.

(*a*)

(*b*)

Fig. 10.9. (*a*) The viscoelastic liquid is poured into a shallow dish filled with the same liquid. (*b*) When the falling stream is thin enough, it rises again following a trajectory which (*f*) *The Kaye effect*

In 1963 Kaye found that if a viscoelastic liquid were poured from a flask into a dish containing the same liquid, then, when the falling stream was thin enough, the stream would bounce off the liquid in the dish every few seconds and rise in an arc before falling back into the dish. This phenomenon is at present unexplained but it makes a very striking demonstration as may be seen in fig. 10.9.

10.7. *Sinusoidal straining of a viscoelastic liquid*

A viscoelastic liquid can be subjected to a sinusoidal stress of small amplitude by using a coaxial cylinder viscometer. The liquid is placed between the two cylinders and the outer cylinder is rotated sinusoidally through a small angle $\phi = \phi_0 \sin \omega t$. This causes the enclosed liquid to oscillate and the movement of the liquid will also cause the suspended inner cylinder to oscillate with the same angular frequency ω. The oscillations of this inner cylinder will however not be in phase with those of the outer cylinder. If this phase difference is δ then $\theta = \theta_0 \sin(\omega t + \delta)$ will represent the oscillations of this inner cylinder.

To explain this phase difference we can argue as follows. If the space between the cylinders were occupied by a perfectly elastic solid the two cylinders would oscillate in phase, that is, $\delta = 0$. At the other extreme, if this space were filled by a purely viscous liquid, the shear stress transmitted to the inner cylinder is

$$\tau = \eta \times (\text{rate of change of shear strain})$$

(see section 9.5). If the sinusoidal shear strain applied to the system by the outer cylinder is

$$\epsilon = \epsilon_0 \sin \omega t \tag{10.2}$$

then

$$\tau = \eta \frac{d\epsilon}{dt}$$

$$= \epsilon_0 \omega \sin\left(\omega t + \frac{\pi}{2}\right). \tag{10.3}$$

From equations (10.2) and (10.3) we see that the shear stress τ on the inner cylinder is $\pi/2$ out of phase with the shear strain applied by the outer cylinder. Thus the two cylinders oscillate with a phase difference of $\pi/2$.

When we have a *viscoelastic* liquid between the cylinders the phase difference δ between the oscillations of the two cylinders will lie between 0 and $\pi/2$ and δ is a measure of the elasticity of the liquid. During each swing a part of the energy is stored elastically as the amplitude increases and then recovered as this amplitude decreases. The remainder of the energy is lost as heat due to the viscous forces and is not recovered.

In the theory of this experiment it is necessary to introduce a *complex* modulus G^* given by

$$G^* = G' + \mathrm{i}G''. \tag{10.4}$$

G' is the real elastic modulus and is a function of the energy stored in the swing; G'' is the 'viscous' modulus which is a function of the energy lost as heat. The phase angle δ is given by

$$\tan \delta = G''/G'.$$

For a purely elastic solid $G''=0$ and $\delta=0$; for a purely viscous liquid $G'=0$ and $\delta=\pi/2$.

The dimensions of a modulus are related to those of viscosity by the relation

$$[G] = \left[\frac{\eta}{\text{time}}\right]$$

(see equation (9.11)); so we will write

$$G'' = \eta''\omega, \text{ etc.}$$

since ω has the dimensions of T^{-1}. Thus equation (10.4) becomes

$$G^* = G' + \mathrm{i}\eta''\omega.$$

Dividing both sides by ω we can rewrite this as

$$\eta^* = \eta' + \mathrm{i}\eta''$$

where η^* is called the *complex dynamic viscosity*.

For further details the reader is referred to Chapter 7 of *Elementary Rheology* by G. W. Scott Blair.

References

(1) Collyer, A. A., 1973 a, *Physics Education*, **8**, 111.
(2) Collyer, A. A., 1973 b, *Physics Education*, **8**, 333.
(3) Collyer, A. A., 1974 a, *Physics Education*, **9**, 38.
(4) Collyer, A. A., 1974 b, *Physics Education*, **9**, 313.
(5) Scott Blair, G. W., *Elementary Rheology* (Academic Press Inc. Ltd London, 1969).
(6) Cheng, D. C.-H., 1973, *Nature*, **245**, 93.

APPENDIX (1)

the ‘ rule of equal areas ’ and the van der Waals isotherm

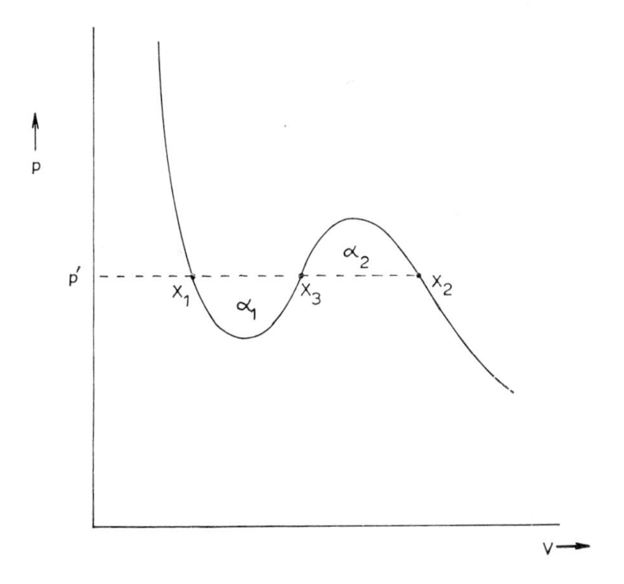

Fig. A.1. The ‘ rule of equal areas ’ and the van der Waals isotherm.

Referring to the above diagram in fig. A.1 choose two points X_1, X_2 as shown such that X_1X_2 is parallel to the V-axis. If the pressure p' at which the change of phase from liquid to vapour in equilibrium with it is to be the value of p at X_1 and X_2 then the position of the line X_1X_2 is determined from three facts:

(1) At equilibrium the Gibbs functions, G, per unit mass of liquid and vapour must be equal, by the general results of thermodynamics.

(2) The thermodynamic relation $(\partial G/\partial p)_T = V$ holds, where V refers to unit mass.

(3) The same relation between p and V holds for the liquid and vapour branches of the curve in fig. A.1.

Thus if G_1, G_2 denote the values of G at X_1, X_2 respectively we can write

$$G_2 - G_1 = \int_{X_1}^{X_2} V\mathrm{d}p$$

$$= \int_{X_1}^{X_3} V\mathrm{d}p + \int_{X_3}^{X_2} V\mathrm{d}p$$

$$= \alpha_1 - \alpha_2 \text{ (using the appropriate sign convention)}$$

where α_1 and α_2 are the areas shown.

Using the fact that $G_1 = G_2$ at equilibrium implies that $\alpha_1 = \alpha_2$, which is the Maxwell 'rule of equal areas'.

APPENDIX (2)

the values of *a* and *b* for a van der Waals gas

The values of a and b for a van der Waals gas can be obtained from the imperfect gas theory and by using the form of $\phi(r)$ appropriate to the van der Waals gas. In section 4.6 we saw that this was

$$\phi(r) = \infty, \quad r < \sigma$$

$$\phi(r) = -\frac{\text{const}}{r^6}, \quad r > \sigma$$

where σ is the diameter of the rigid spherical molecule. For the detailed calculations the reader is referred to Rushbrooke's "Statistical Mechanics", Chapter 16. The value of a for 1 mole turns out to be given by

$$a = -2\pi N_A^2 \int_\sigma^\infty r^2 \phi(r) \mathrm{d}r$$

where N_A is the number of molecules in a mole.

The value of b is given by

$$b = 2\pi N_A \int_0^\sigma r^2 \mathrm{d}r = \frac{2}{3} N\pi\sigma^3 = 4Nv_0$$

where

$$v_0 = \frac{4\pi}{3}\left(\frac{\sigma}{2}\right)^3$$

s the volume of one of the rigid spherical molecules.

FURTHER READING

BOOKS

BARTON, A. F. M., 1974, *The Dynamic Liquid State* (Longman).

COLE, G. H. A., 1967, *An Introduction to the Statistical Theory of Classical Simple Dense Fluids* (Pergamon Press).

DAVIES, Mansel, 1965, *Molecular Behaviour*, Chapter 5 (Pergamon Press).

EGELSTAFF, P. A., 1967, *An Introduction to the Liquid State* (Academic Press, London and New York).

FISHER, I. Z., 1964, *Statistical Theory of Liquids* (The University of Chicago Press, Chicago and London).

HAMANN, S. D., 1957, *Physico-Chemical Effects of Pressure* (Butterworths Scientific Publications, London).

HIRSCHFELDER, J. O., CURTISS, C. F., and BIRD, R. B., 1954, *Molecular Theory of Gases and Liquids* (Wiley, New York).

LODGE, A. S., 1964, *Elastic Liquids*, Chapter 10 (Academic Press).

MEARES, P., 1965, *Polymers*, Chapter 11 (D. van Nostrand Co. Ltd., London).

RICE, S. A., and GRAY, P., 1965, *The Statistical Mechanics of Simple Liquids*, Chapter 2 (Interscience Publishers).

ROWLINSON, J. S., 1959, *Liquids and Liquid Mixtures* (Butterworths Scientific Publications, London).

RUSHBROOKE, G. S., 1949, *Statistical Mechanics* (Oxford University Press).

TEMPERLEY, H. N. V., ROWLINSON, J. S., and RUSHBROOKE, G. S. (Editors), 1968, *Physics of Simple Liquids* (North-Holland Publishing Company, Amsterdam).

UBBELOHDE, A. R., 1965, *Melting and Crystal Structure* (Oxford University Press).

WALTERS, K., 1975, *Rheometry* (Chapman and Hall).

PAPERS

BARKER, J. A., and HENDERSON, D., 1968, 'The Equation of State of Simple Liquids', *J. Chem. Educ.*, **45,** 2.

ENDERBY, J. E., 'Neutron Scattering in Liquids' (This is Chapter 14 of *Physics of Simple Liquids* referred to in the above previous list of book references).

HAYWARD, A. T. J., 1964, 'Measuring the Extensibility of Liquids', *Nature*, **202,** 481.

HAYWARD, A. T. J., 1970, 'New Laws for Liquids: Don't Snap, Stretch!', *New Scientist*, **45,** 196.

POWLES, J. G., 1974, 'Liquids—The Awkward In-between', *Contemp. Phys.*, **15,** 409.

REES, E. P., and TREVENA, D. H., 1966, 'A Study of the Berthelot Method of Measuring Tensions in Liquids', *Brit. J. Appl. Phys.*, **17,** 671.

TABOR, D., 1964, 'Large-scale Properties of Matter', *Contemp. Phys.*, **6,** 112.

WOODHEAD-GALLOWAY, J., 1972, 'Towards the Structure of Liquids', *New Scientist*, **56,** 399.

INDEX